# APPLICATION OF OMICS, AI AND BLOCKCHAIN IN BIOINFORMATICS RESEARCH

# ADVANCED SERIES IN ELECTRICAL AND COMPUTER ENGINEERING

Editors: W.-K. Chen *(University of Illinois, Chicago, USA)*
Y.-F. Huang *(University of Notre Dame, USA)*

The purpose of this series is to publish work of high quality by authors who are experts in their respective areas of electrical and computer engineering. Each volume contains the state-of-the-art coverage of a particular area, with emphasis throughout on practical applications. Sufficient introductory materials will ensure that a graduate and a professional engineer with some basic knowledge can benefit from it.

*Published:*

Vol. 21:   *Application of Omics, AI and Blockchain in Bioinformatics Research*
edited by Jeffrey J. P. Tsai and Ka-Lok Ng

Vol. 20:   *Computational Methods with Applications in Bioinformatics Analysis*
edited by Jeffrey J. P. Tsai and Ka-Lok Ng

Vol. 18:   *Broadband Matching: Theory and Implementations (Third Edition)*
by Wai-Kai Chen

Vol. 17:   *Practical Signal Processing and Its Applications:*
*With Solved Homework Problems*
by Sharad R Laxpati and Vladimir Goncharoff

Vol. 16:   *Design Techniques for Integrated CMOS Class-D Audio Amplifiers*
by Adrian I. Colli-Menchi, Miguel A. Rojas-Gonzalez and
Edgar Sanchez-Sinencio

Vol. 15:   *Active Network Analysis: Feedback Amplifier Theory (Second Edition)*
by Wai-Kai Chen (University of Illinois, Chicago, USA)

Vol. 14:   *Linear Parameter-Varying System Identification:*
*New Developments and Trends*
by Paulo Lopes dos Santos, Teresa Paula Azevedo Perdicoúlis,
Carlo Novara, Jose A. Ramos and Daniel E. Rivera

Vol. 13:   *Semiconductor Manufacturing Technology*
by C. S. Yoo

Vol. 12:   *Protocol Conformance Testing Using Unique Input/Output Sequences*
by X. Sun, C. Feng, Y. Shen and F. Lombardi

Vol. 11:   *Systems and Control: An Introduction to Linear, Sampled and*
*Nonlinear Systems*
by T. Dougherty

For the complete list of titles in this series, please visit
http://www.worldscientific.com/series/asece

Advanced Series in Electrical and Computer Engineering – Vol. 21

# APPLICATION OF OMICS, AI AND BLOCKCHAIN IN BIOINFORMATICS RESEARCH

### Editors

## Jeffrey J.-P. Tsai
Asia University, Taiwan

## Ka-Lok Ng
Asia University, Taiwan

W World Scientific

NEW JERSEY · LONDON · SINGAPORE · BEIJING · SHANGHAI · HONG KONG · TAIPEI · CHENNAI · TOKYO

*Published by*

World Scientific Publishing Co. Pte. Ltd.

5 Toh Tuck Link, Singapore 596224

*USA office:* 27 Warren Street, Suite 401-402, Hackensack, NJ 07601

*UK office:* 57 Shelton Street, Covent Garden, London WC2H 9HE

**Library of Congress Cataloging-in-Publication Data**

Names: Tsai, Jeffrey J.-P., editor. | Ng, Ka-Lok, editor.

Title: Application of omics, AI and blockchain in bioinformatics research / edited by
    Jeffrey J-P Tsai, Asia University, Taiwan and Ka-Lok Ng, Asia University, Taiwan.

Description: New Jersey : World Scientific, [2019] | Series: Advanced series in electrical and
    computer engineering ; volume 21 | Includes bibliographical references and index.

Identifiers: LCCN 2019008597 | ISBN 9789811203572 (hc : alk. paper)

Subjects: LCSH: Life sciences--Data processing. | Biotechnology--Data processing. |
    Biomedical engineering--Data processing. | Bioinformatics. | Blockchains (Databases) |
    Artificial intelligence.

Classification: LCC QH324.2 .A66 2019 | DDC 610.285--dc23

LC record available at https://lccn.loc.gov/2019008597

**British Library Cataloguing-in-Publication Data**

A catalogue record for this book is available from the British Library.

For any available supplementary material, please visit
https://www.worldscientific.com/worldscibooks/10.1142/11364#t=suppl

Desk Editors: Herbert Moses/Yu Shan Tay

Typeset by Stallion Press
Email: enquiries@stallionpress.com

# Preface

Biomedical data are complex in nature; they are composed of multi-level data that range from the molecular level to the network level to the cell level then to the physiological level. A major challenge is to combine multi-level data or omics data in such a way that we can gain useful knowledge from them. To attack these problems, both machine learning and deep learning techniques, two of the subfields in artificial intelligence, emerge as highly effective approaches to address complex data analysis problems. As the volume of biomedical data increases, so does the demand for establishing and building the big data analytic approach for solving a wide range of applications.

This monograph is a follow-up of our previous volume entitled *Computational Methods with Applications in Bioinformatics Analysis*. It is composed of 10 chapters and presents the use of different computational methods, such as deep learning, big data analysis, advance computational approach, and network analysis technique to address problems in precision medicine.

This book is a collection of 10 chapters written by international researchers with expertise in omics data analysis, the use of AI in autism case study, blockchain application for clinical platform, circulating tumor DNA, big data computation of drug design, cluster analysis of RNA-seq data, cancer gene signatures, tandem mass spectrometry, Boolean models of metabolic networks, and tensor decomposition-based unsupervised feature selection.

In short, this book presents the use of AI, blockchain, and big data analysis for omics study and a range of biomedical applications. Instead of focusing on problems in theoretical biology modeling, the book provides an in-depth treatment of a wide range of biomedical issues.

Jeffrey J.P. Tsai
Ka-Lok Ng
Taichung, 2019

# About the Editors

**Jeffrey J.P. Tsai** received a Ph.D. degree in Computer Science from the Northwestern University, Evanston, Illinois. He is currently the President of Asia University, Taiwan.

Dr. Tsai was a Professor of Computer Science and the Director of the Distributed Real-Time Intelligent Systems Laboratory at the University of Illinois, Chicago. He was also an Adjunct Professor at Tulane University, a Visiting Professor at Stanford University, a Visiting Scholar at the University of California at Berkeley, and a Senior Research Fellow of $IC^2$ at the University of Texas at Austin.

His research interests include bioinformatics, big data, distributed real-time systems, knowledge-based software engineering, formal modeling and verification. His research has been supported by United States (US) National Science Foundation (NSF), Defense Advanced Research Projects Agency (DARPA), United States Air Force (USAF) Rome Laboratory, Department of Defense, Army Research Laboratory, Motorola, Fujitsu, and Gtech. The technology on knowledge-based software engineering developed by him and his research team resulted in the world-first complete transformation of an embedded software product in 1993 and is now used to produce communication software systems worldwide. Tsai coauthored *Knowledge-Based Software Development for Real-Time Distributed Systems* (World Scientific, 1993), *Distributed Real-Time Systems* (Wiley, 1996), *Compositional Verification of Concurrent and Real-Time Systems* (Springer/Kluwer, 2002), *Security Modeling and Analysis of Mobile Agent Systems* (Imperial College Press, 2006), *Intrusion Detection: A Machine Learning Approach* (Imperial College Press, 2010), and coedited *Monitoring and Debugging of Distributed Real-Time Systems* (IEEE Computer Society Press, 1995), *Machine Learning Applications in Software Engineering* (World Scientific, 2005), *Ubiquitous Intelligence and Computing* (Springer, 2006), and *Machine Learning in Cyber Trust: Security, Privacy, Reliability* (Springer, 2009).

Tsai was the Conference Co-Chair of the 16th IEEE International Symposium on Software Reliability Engineering, the 9th IEEE International Symposium on Multimedia, the 1st IEEE International Conference on Sensor Networks, Ubiquitous and Trustworthy Computing, and the 3rd IFIP International Conference on Ubiquitous Intelligence and Computing. From 2000 to 2003, Dr. Tsai chaired the IEEE/CS Technical Committee on Multimedia Computing and served on the steering committee of the IEEE Transactions on Multimedia. He was an Associate Editor of the *IEEE Transactions on Knowledge and Data Engineering*, and *IEEE Transactions on Services Computing*. He is also the co-Editor-in-Chief of the *International Journal on Artificial Intelligence Tools* and book series on Health Informatics.

Dr. Tsai has served on the IEEE Distinguished Speaker program, US DARPA Information Science and Technology (ISAT) Working Group, and on the review panel for US NSF and National Institutes of Health (NIH). He received an Engineering Foundation Research Award from the IEEE and the Engineering Foundation Society, a University Scholar Award from the University of Illinois Foundation, an IEEE Technical Achievement Award and an IEEE Meritorious Service Award from the IEEE Computer Society. He is a Fellow of the American Association for the Advancement of Science (AAAS), the Institute of Electrical and Electronics Engineers (IEEE), and the Society for Design and Process Science (SDPS).

**Ka-Lok Ng** received the Honors diploma in Physics from Hong Kong Baptist College in 1983, and the Ph.D. degree in theoretical elementary particle physics from the Vanderbilt University, US, in 1990. He is currently the Distinguished Professor at the Department of Bioinformatics and Medical Engineering, Asia University, Taiwan. Beginning from December 2009, he served on the Editorial Board of several international journals. He was the Editor-in-Chief (2010–2014) of the World Scientific and Engineering Academy and Society (WSEAS) Transactions of Biology and Biomedicine.

He was the Conference Program Co-Chair of the IEEE International Conference on BioInformatics and BioEngineering and IAENG International Conference of Bioinformatics. Dr. Ng has served on the program committee of several well-known bioinformatics conferences: International Conference on Genome Informatics (GIW), Asia-Pacific Bioinformatics Conference (APBC), and International Conference on Bioinformatics (InCoB). He has served on the invited speaker program of the International Association of Engineers (IAENG) International Conference of Bioinformatics since the last 4 years.

Furthermore, Dr. Ng is also actively involved in reviewing manuscripts for international leading journals such as *Bioinformatics* and *Nature Scientific Reports*.

His research has been supported by the Ministry of Science and Technology of Taiwan, Asia University and China Medical University.

Dr. Ng publishes articles in highly ranked journals, in the areas of systems biology, drug repositioning, role of epigenetics in cancer biology, protein function prediction, and DNA data-hiding method. His research interest include system biology, cancer biology, multi-omics data analysis, and cosmology.

# Acknowledgment

We are grateful to all the authors for their efforts and involvement in producing this book. This book would not have been possible without the financial support by the Ministry of Science and Technology of Taiwan (MOST) under the grant number MOST 107-2632-E-468-002, and support from Asia University. Finally, we would like to thank the editorial and production staffs at World Scientific Publishing for making this book possible.

# Contents

*Preface*                                                                  v

*About the Editors*                                                      vii

*Acknowledgment*                                                         xi


Chapter 1.  Generalized Iterative Modeling for Clinical
            Omics Data Analysis                                           1

            *Kung-Hao Liang*


Chapter 2.  Explainable AI: Mining of Genotype Data
            Identifies Complex Disease Pathways —
            Autism Case Studies                                          11

            *Matt Spencer, Saad Khan, Zohreh Talebizadeh,
            and Chi-Ren Shyu*


Chapter 3.  Blockchain for Pre-clinical and Clinical Platform
            with Big Data                                                 29

            *Yin-Wu Chen and Zon-Yin Shae*


Chapter 4.  Analysis of Circulating Tumor DNA in Patients
            with Cancer: A Clinical Perspective                          47

            *Chi-Chun Yeh and Peter Mu-Hsin Chang*

Chapter 5.  Big Data Computation of Drug Design:
From the Natural Products to the
Transcriptomic-Based Molecular Development        59

*David Agustriawan, Arli Aditya Parikesit,
and Rizky Nurdiansyah*

Chapter 6.  A Hybrid Approach Integrating Model-Based
Method and Gene Functional Similarity
for Cluster Analysis of RNA-Seq Data             87

*Ming-Han Chan, Pin-Chen Chou,
Rong-Ming Chen, and Rouh-Mei Hu*

Chapter 7.  High-Performance Computing for Measurement
of Cancer Gene Signatures                        109

*Hsueh-Ting Chu*

Chapter 8.  High-Performance Computing in Tandem Mass
Spectrometry (MS/MS) Data Processing             123

*Li Chuang and Lin Feng*

Chapter 9.  Analysis of Boolean Networks and Boolean
Models of Metabolic Networks                     141

*Tatsuya Akutsu*

Chapter 10. Tensor Decomposition Based Unsupervised
Feature Extraction Applied to Bioinformatics     159

*Y-h. Taguchi*

*Index*                                          189

Chapter 1

# Generalized Iterative Modeling for Clinical Omics Data Analysis

Kung-Hao Liang

*Department of Medical Research,*
*Taipei Veterans General Hospital, Taiwan*
*Institute of Food Safety and Health Risk Assessment,*
*National Yang-Ming University, Taiwan*
*Institute of Biomedical Informatics,*
*National Yang-Ming University, Taiwan*

## Abstract

Artificial intelligence has shown great potential in many aspects of human life, including the biomedical research and clinical care. Methods such as artificial neural networks are capable of handling large volumes of data from various medical image modalities, or from "omics" technologies, including genomics, transcriptomics, proteomics, metabolomics, and glycomics. However, these technologies often offered "black box" solutions where the expressibility of these numerical models were not satisfactory. Here, we developed the generalized iterative modeling (GIM) method, extending the conventional generalized linear models with a machine-learning twist. This method features an iterative shaping of highly-expressive polynominal models with automatically determined combinations of clinical and omics variables. The models can be written in a few lines of mathematical equations, allowing human comprehensions and interpretations. It will also facilitate the implementations in a wide-diversity of hardware and software platforms. This GIM software is now available for optimizing the U-statistics, F-statistics, and the log-likelihood in the Cox proportional hazards model. Using real data, the performance of GIM was demonstrated to be better than those from the generalized linear models and the orthogonal partial least squares discriminant analysis. The source code of GIM can be found at the GitHub site https://github.com/khliang/GIM.

*Keywords*: artificial intelligence, model expressibility, metabolomics, survival analysis, Cox regression.

## 1.1.  Introduction

An important goal of clinical investigations is to clarify the underlying quantitative relationships between clinical variables. The relationships not only illuminates the mechanism of human in health and disease, but also facilitates clinical diagnosis and outcome prediction. An explicit mathematical equation serves an universal language for communication between clinicians and scientists beyond country borders and across generations. The generalized linear models are a family of methods frequently used for multivariate clinical investigations. They include the logistic regression which is useful for classifying dichotomous states (e.g. diseased and healthy states) using a linear combination of variables; the analysis of variance (ANOVA) which is often used for analyzing three or more states; the repeat-measure ANOVA or generalized estimating equations which are used for repeatedly collected time-course data; and the Cox proportional hazards models which are often used in longitudinal time-to-event analysis.

Variables in clinical investigations include, for example, the status or risk of a disease, the serum biochemistry measurements, the urine measurements, the biomedical imaging features, as well as the basic variables such as age and gender. In recent decades, a major driving force of biomedical science was the advancement of the "omics" technology, including genomics, transcriptomics, proteomics, metabolomics, and glycomics. Using the clinical samples (blood, urine, tissue, cerebrospinal fluids), these technologies can generate systematic and comprehensive measurements. However, variables generated from these technologies are often in the order of hundreds or thousands, which create enormous challenges for conventional methods such as the generalized linear models. The artificial intelligence (AI) technology offers a promise of handling data with excessive volume and diversity. One representative class of AI technology is the artificial neural networks, which have various modern architectures such as convolutional neural network, recurrent neural network, generative adversarial network, to name a few. They have recently achieved many milestones such as surpassing the human capabilities in playing the game of Go, achieving high accuracy in image recognition, and offering products and services utilizing human voice recognition. Basically, artificial neural networks are used for multivariate analysis. An effective integration of AI and omics technologies can offer unprecedented values in biomedical science and clinical care.

However, the artificial neural networks often provided "black box" solutions which lacked expressibility. If the "black boxes" were to be opened, we would find the solutions in the formats of multiple layers of hidden nodes with enormous amount of state parameters connected by complex wirings. This is difficult for our interpretation and intuitive understanding. Despite their excellent empirical performance, the lack of expressibility prevented them from being precisely documented in medical literature.

The generalized linear models have excellent expressibility consisting of three specific components: (1) a linear combination of variables, denoted as M; (2) a link function connecting M to the response variable Y; and (3) a goal of optimization such as the log-likelihood. In (1), not every variables assessed by the "omics" are actually associated to the response variable. Thus, data-driven regularization methods such as the Lasso regression methods, or the elastic net methods were proposed which can achieve variable selection by reducing the coefficients of irrelevant variables toward 0.

Conventionally, multivariate generalized linear models were used for analyzing clinical variables and laboratory measurements. This method can only provide the weighted addition of variables. Occasionally, multiplicative interactive variables were pre-defined *a priori* as a new compound variable, and analyzed in the same way as other variables. Unfortunately, most of the time, the interactive relationships between variables were unknown and were simply ignored. The number of potential interactions among variables escalate as the number of variables increases. An exhaustive manual search of all possible variable combinations is prohibitive due to the factorial explosion nature of all combinations, particularly when the number of variables is large in "omics" studies. Data-driven approach for the identification of novel variable interactions are thus urgently required.

With such a motivation in mind, we introduced the generalized iterative modeling (GIM) framework with an emphasis of expressibility in results, and the capability of numerical learning as in the AI technology. The component (1) of the conventional models were relaxed to incorporate all polynomial combination of variables, enabling the assessment of the conditional or synergistic effects of the variables. GIM is a generalization of the previously developed GABA [1] (for genotype analysis) and HABA [2] (for haplotype analysis) algorithm (Liang *et al.*, 2006, 2007), which were originally restricted to Boolean algebra due to the discrete nature of genetic variables. Consistent with previous algorithms, the solver employed the genetic programming approach. This chapter was organized as follows. Section 1.2 described the GIM framework. Section 1.3 showed the applications of GIM to real biomedical data. We demonstrated that GIM outperformed conventional Cox regression for time-to-event analysis, and the logistic regression for biomedical state classification.

## 1.2. Methods

### 1.2.1. *A Polynomial Combination of Variables in GIM*

Similar to the generalized linear models, GIM also has three important components: (1) a polynomial combination of variables, denoted as $M$; (2) a link function connecting $M$ to the response variable $Y$; and (3) a goal of optimization such as the likelihood ratio. $M$ comprises subset of variables $V(V1, V2, V3...)$ joined together by algebraic operations (i.e. addition "$+$" and multiplication "$*$"), and weighted by coefficients ($C1, C2, C3...$). For example, one model may appear like this

$$M(t) = C1V1 + C2V2 * C3V3 - C4V4 * C5V5$$

The multiplication operation is the major difference of GIM to conventional generalized linear models. This way, GIM can address the nonlinear effect of conditioning and synergistic effects.

### 1.2.2. *Optimization Targets*

Optimization targets of GIM include the following: (1) The log-likelihood of the proportional hazards models for the time-to-event analysis. (2) The non-parametric U-statistics, which equates to the area under the receiver operating characteristic (ROC) curve, for the classification of the dichotomous states. (3) The F-statistics (the ratio of between class variance to the within class variance) for the classification of three or model states. All these optimization targets have been implemented in the current code written in the C programming language.

### 1.2.3. *The Genetic Programming as the Optimizer for GIM*

The GIM method features an iterative modeling strategy where models (with addition and multiplication operators linking the selected variables) were progressively shaped for the fitting of high volumes of clinical and genomic data. This is the concept of adaptation which is analogous to "learning" in the AI technology.

The adaptive plan determines the model at time $(t + 1)$ based on the structural modifications of model at time $(t)$ in response to the environment. The discrete time interval $(t)$ is used here for the simplicity of method description. There are literally infinite numbers of combinations among even dozens of variables, therefore, the strategy behind the algorithm was to generate and sculpture models progressively with improved performance. The adaptive plan is defined by the following computational operations:

*Coefficient adjustment*

$$\text{Example: } M(t + 1) = C6V1 + C2V2 * C3V3 - C4V4 * C5V5$$

*Adding or removing clinical variables*

$$\text{Example: } M(t + 2) = C6V1 + C2V2 * C3V3$$

*Changing the algebraic operators between variables*

$$\text{Example: } M(t + 3) = C6V1 * C2V2 + C3V3$$

*And a combination of parts of two models*

$$\text{Example: } M(t + 4) = C7V7 + C8V8 * C9V9 * C2V2 + C3V3$$

In this algorithm, new models can replace original candidate models as new candidates if they can achieve better performance. Otherwise, the original candidate models were retained and continued to be analyzed. The whole process was iterated until quasi-optimal predictors were found when the performance does not increase further after a predefined number of iterations.

## 1.3.  Results

### 1.3.1.  *GIM for the Time-to-Event Analysis on Metabolomics Data*

Cox regression is one of the most widely used method for longitudinal time-to-event analysis in biomedical literature. In this application, we used both the GIM and Cox regression methods for analyzing the same time-to-event data with baseline "omics" measurements. This is a prospectively recruited cohort of 253 hepatitis B-related liver cirrhotic patients, and the observatory event is the occurrence of hepatocellular carcinoma (HCC). A total of 14 patients were observed to develop HCC during the follow-up period. The baseline measurements are serum metabolomics data measured by the nuclear magnetic resonance spectrometry. The goal was to find a model comprising baseline variables which can predict the occurrence of HCC, i.e. a typical time-to-event analysis in the medical literature.

Among all the 830 metabolite measurements, a total of 34 variables showed significant difference (univariate $t$-test $P < 0.05$) between those with HCC and those without HCC in the follow-up time. The multivariate analysis was then performed by GIM on these variables, and a model was derived as:

The hazard function of the patient $i$:

$$H(t|Score\_i) = H_0(t)\exp(Score\_i)$$

where
$$Score\_i =$$
$$(-17.8659) * biomarker1\_i$$
$$+(122.3357) * biomarker3\_i$$
$$-(43.2095) * biomarker6\_i$$
$$+(42.5202) * biomarker8\_i$$
$$+(21.8199) * biomarker24\_i$$
$$+(3851.1388) * biomarker23\_i * biomarker7\_i$$

$$- (229.6776) * (\text{biomarker1\_}i)^2 * \text{biomarker24\_}i$$
$$- (60118.0411) * (\text{biomarker33\_}i)^4$$

There are eight biomarkers (i.e. metabolite measurements) selected automatically by GIM into this model. We then performed the conventional Cox regression on the same variables using the SPSS software. The achieved $-2 \times$ log-likelihood is 154.185. In contrast, the $-2 \times$ log-likelihood achieved by the GIM is 116.49, a much lower value than that achieved by conventional Cox regression.

The patients were then stratified into 10 groups based on their HCC risk scores given by the GIM and the conventional Cox regression. The cumulative incidence of HCC were shown using the Kaplan–Meier plots (Fig. 1.1(a)). The highest risk patient strata (i.e. the 10th decile), and the second highest risk patient strata (i.e. the 9th decile) both have significantly

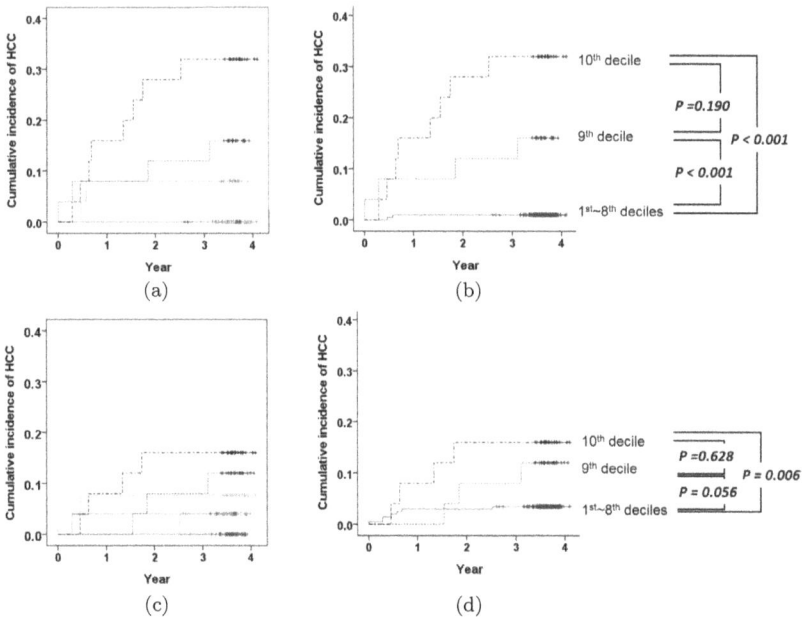

Fig. 1.1. The Kaplan–Meier plots of strata of patients whose HCC risks were scored by GIM and the conventional Cox regression. (a) 10-decile patient strata when patients were scored by GIM. (b) The result of (a) where decile 1st–8th were merged together. (c) 10-decile patient strata when patients were scored by the Cox regression. (d) The result of (c) where decile 1st–8th were merged.

higher cumulative incidence than the rest of the patients (i.e. 1st–8th decile, the low risk group) when the GIM was used ($P < 0.001$ respectively, Fig. 1.1(b)). In comparison, only the highest risk patient strata ($P = 0.006$) but not the second highest risk patient strata ($P = 0.056$) has higher cumulative incidence than the low-risk patients when the conventional Cox regression was used (Figs. 1.1(c) and (d)). Hence, this analysis demonstrate the superiority of GIM to conventional Cox regression.

### 1.3.2.  *GIM for the Classification of Liver Fibrosis Patients*

Liver fibrosis hampers liver function and also raises the risk of hepatocellular carcinoma. As a result, knowing the stage of liver fibrosis is critical for the clinical management of patients. Currently, the standard staging method is by the pathologists' evaluation on the liver biopsy specimens, according to the Ishak fibrosis staging system. To reduce the pathologist-to-pathologist scoring variability, we have previously reported an automatic liver fibrosis staging system, based on the second harmonic microscopy technology and an automatic algorithm, for scanning automatically the biopsy specimens, extract image features and then perform a staging analysis using the

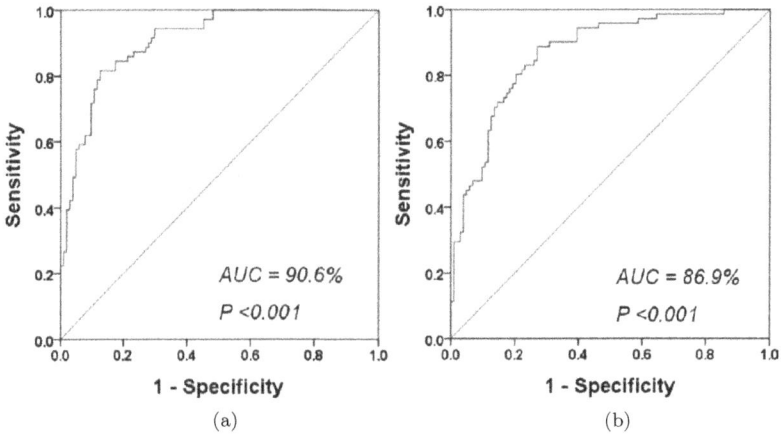

Fig. 1.2.   Performance comparison of GIM versus conventional logistic regression (calculated using the SPSS software) in the analysis of second harmonic microscopy data. The AUC can be improved from 86.9% (by conventional logistic regression) to 90.6% (by GIM).

image features [3]. The data was used here to evaluate the newly developed GIM algorithm and compare its performance with the conventional logistic regression methods. The 13 image features remained being used here [3]. When classifying patients to either prominent fibrosis (Ishak stage 4–6 scored by pathologist) or early fibrosis (Ishak stage 1–3), the conventional logistic regression method achieved an AUC of 86.9% (Fig. 1.2(b)).

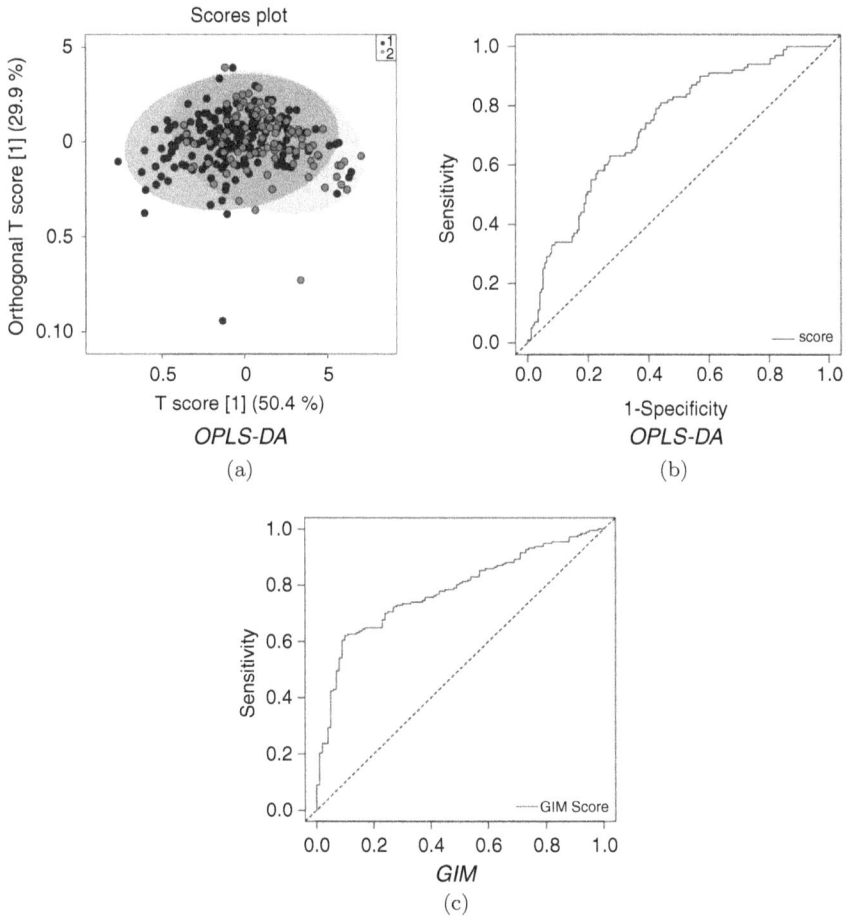

Fig. 1.3. Cross-sectional classifications of HBV and HCV patients using two different algorithms. (a) The scatter plot produced by the OPLS-DA algorithm; (b) the receiver operating characteristic curve of the OPLS-DA performance (AUC = 73.3%, $P < 0.001$); (c) the curve of the GIM performance (AUC = 77.9%, $P < 0.001$).

Despite the already good performance of conventional logistic regression, the GIM algorithm can still achieve a better performance with an AUC of 90.6% (Fig. 1.2(a)).

### 1.3.3. GIM has Superior Performance in Classification than the OPLS-DA Method

The Orthogonal Projections to Latent Structures Discriminant Analysis (OPLS-DA) is a widely used multivariate analysis method particularly in the field of metabolomics. In this analysis, OPLS-DA and GIM were both applied to the same sets of metabolomics data. It turned out that GIM outperformed the OPLS-DA in the classification of chronic hepatitis B (HBV) and C (HCV) patients, based on their serum metabolomics profiles (Fig. 1.3).

### 1.4.  Conclusions

In this chapter, we presented the generalized iterative modeling method which is an augmentation of conventional generalized linear models, a widely used method with high expressibility. We demonstrated the capability of GIM in time-to-event analyses and state classifications, which are cornerstone to clinical investigations. Empirical evaluations showed that the performance of GIM is better than those of conventional methods.

### References

1.  Liang, K.-H., Hwang, Y., Shao, W.-C. and Chen, E. Y. (2006). An algorithm for model construction and its applications to pharmacogenomic studies. *Journal of Human Genetics* 51, pp. 751–759.
2.  Liang, K.-H. and Wu, Y.-J. (2007). Prediction of complex traits based on the epistasis of multiple haplotypes. *Journal of Human Genetics* 52, pp. 456–463.
3.  Wang, T.-H., Chen, T.-C., Teng, X., Liang, K.-H. and Yeh, C.-T. (2015). Automated biphasic morphological assessment of hepatitis B-related liver fibrosis using second harmonic generation microscopy. *Scientific Reports* 5, p. 12962.

Chapter 2

# Explainable AI: Mining of Genotype Data Identifies Complex Disease Pathways — Autism Case Studies

Matt Spencer*, Saad Khan*, Zohreh Talebizadeh†, and Chi-Ren Shyu*

*Informatics Institute, University of Missouri, Columbia
†Children's Mercy Hospital, Kansas City, MO, USA

## 2.1.  Introduction

Many biologically and clinically relevant questions rely on the identification of implicated genes, using a variety of genomics and bioinformatics methods. Genomes of individuals affected by a disease could be examined for rare mutations, thereafter identifying the genes containing these mutations [1]. Gene expression studies identify lists of genes which have different levels of transcription in diseased cells [2]. Advanced computational techniques are emerging as key contributors to biological and medical research, but in these fields it is critical to obtain results that are explainable and actionable. The results of large-scale artificial intelligence (AI) and computing studies must be obtained in a way which facilitates the understanding of the role of genes in the development of diseases. The results should achieve a level beyond the identification of risk genes and so need to be further analyzed to identify the cellular pathways which affect the progression of the disease. The ultimate goal is to make the computational results actionable by incorporating investigation of these pathways into laboratory and clinical studies.

The push for explainable, actionable results is particularly important during the study of complex disease etiology. In complex genetic disorders like autism [3], cancer [4], and diabetes [5], disease development often cannot be explained by variations in individual genes or pathways [6]. Methods have been employed to identify genes that have some association with these disorders, but they have had limited success explaining disease mechanisms [7, 24]. These fields are ripe for the application of AI algorithms designed to turn mountains of genomic and phenotypic data into meaningful information relevant to understanding the diseases.

### 2.1.1.  *Autism Case Studies*

The tools described in this chapter were developed to address challenges in autism research, though they are similarly applicable for studying other complex diseases. There are three primary attributes that a complex disease should have to fully take advantage of this pipeline. First, it should be established that common genetic variation is a major contributor to disease development. In the case of autism, it is estimated that around half of the genetic contribution to autism stems from common variation, and increasing our understanding of this component of autism etiology is essential [8]. Second, the disease should be complex enough that multiple combinations of interacting genotypes could plausibly contribute to the disease etiology. Diseases which primarily develop due to a single genetic anomaly, or diseases with a lack of phenotypic heterogeneity, may be better served by a methodology which focuses on a few driver genes with large effects. Finally, the disease should be one suspected to comprise multiple subgroups of disease patients, as contrasting distinct disease subgroups is one of the greatest strengths of this procedure. Autism is broadly agreed to be a collection of multiple subgroups, many of which likely have distinct genetic etiologies [9]. This makes it an ideal candidate for the subgroup contrast procedure described as follows.

Since autism effectively demonstrates the impact of the techniques in this chapter, case studies were applied to analyze the Simons Simplex Collection (SSC) — a cohort of 2,600 families with one person diagnosed with autism, collected using the support of the Simons Foundation Autism

Research Initiative [10]. The data include extensive genotypes along with supplementary phenotype attributes describing 11,560 individuals and pedigrees connecting autism patients to their healthy parents and siblings.

Previous exploration of the SSC examined hundreds of potential subgroup-defining phenotypes, highlighting several subgroups which showed signs of distinct genetic etiologies when compared to other autism patients. Many of the included subgroups fell into broad categories, notably groups defined by mild levels of emotional and behavioral problems and groups defined by poorly developing language skills. We selected one promising subgroup from each of these categories for these autism case studies. The first case focused on the Somatic Complaints subgroup of 380 autism patients with low to moderate levels of unwarranted physical maladies (tiredness, headaches, etc., measured by the Child Behavior Checklist for ages 6–18 [1]), comparing this group to 2,177 autism patients with higher levels of these maladies. Similarly, the second case examined the Small Vocabulary subgroup of 1,680 patients who utilized few words in conversation at the time of diagnosis was compared to the 295 patients with average to large vocabularies.

### 2.1.2. *Overview*

This chapter describes an analysis pipeline which transforms raw genotype and phenotype data into knowledge of the functional pathways which play a major role in the development of a complex disease. Shown in Fig. 2.1, a raw data bank of genotypes and phenotypes is explored by Heritable Genotype Contrast Mining (HGCM), a data mining procedure which identifies gene pairs associated with a disease population. A subsequent multi-level pathway enrichment analysis then utilizes three statistical tests to interpret the identified genes as functional pathways relevant to disease development. Finally, a prioritization function ranks the pathways by aggregating the results of the various analyses, guiding future research into the specific contribution of pathways to the development of complex diseases. Throughout this process, we discuss the methodology and impacts of applying these processes to case studies utilizing an autism dataset.

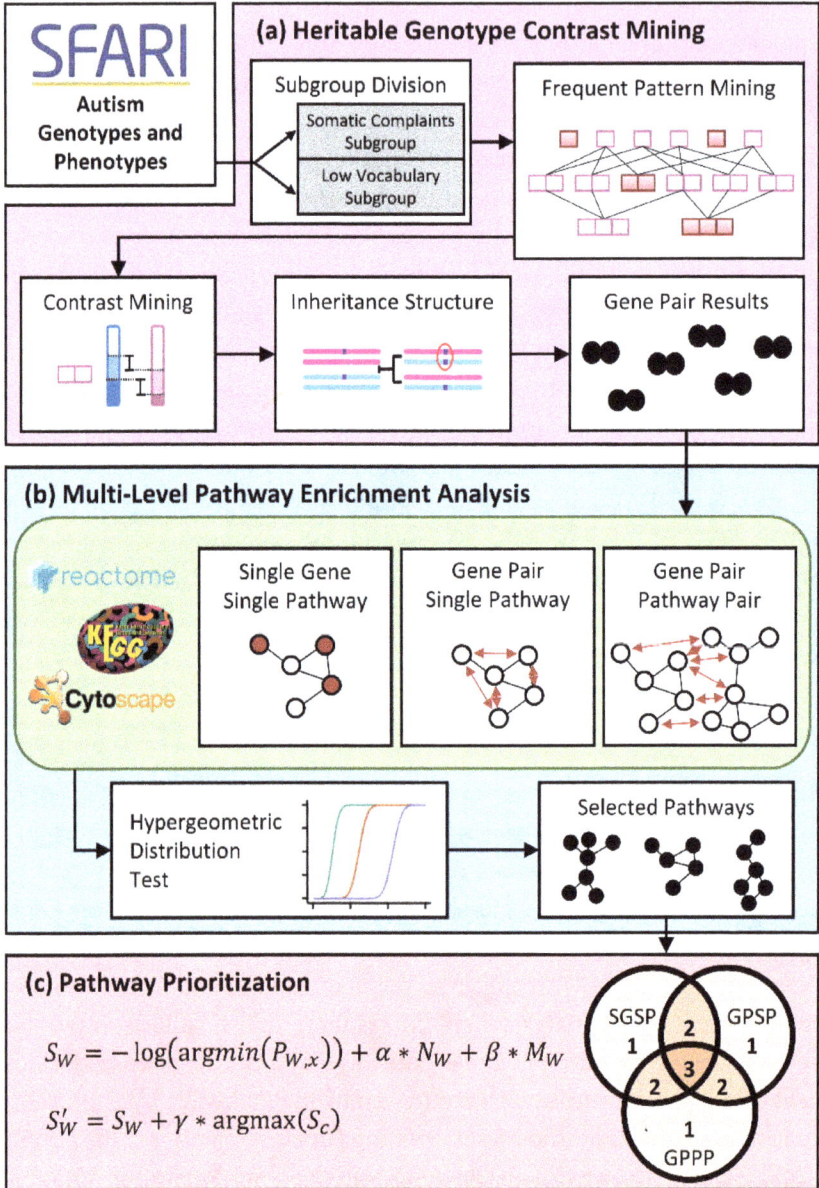

Fig. 2.1.    Flowchart describing the full analysis pipeline. (a) Heritable Genotype Contrast Mining is used to analyze genotypes and phenotypes to highlight gene pairs. (b) Multi-level pathway enrichment analysis is performed, identifying the pathways relevant to the list of gene pairs. (c) Pathways are prioritized for closer examination, accounting for the number of enrichment analyses selecting each pathway.

## 2.2. Heritable Genotype Contrast Mining

HGCM is a procedure for identifying genetic patterns relevant to a specific population of individuals affected by a disease or disorder [12]. Genetic variants measured by single nucleotide polymorphisms are organized into patterns of two or more genetic variants, and the prevalence of each pattern is measured in two compared populations to identify patterns which distinguish the groups. Patterns are then evaluated to determine whether they are unique to the affected individual within their nuclear family. This method outputs selected combinations of genotypes, which can easily be processed into a list of gene pairs appropriate for this pathway enrichment pipeline. The following sections explain these steps in more detail.

### 2.2.1. *Frequent Pattern Mining*

Frequent Pattern Mining (FPM) is a data mining technique that excels at identifying combinations of features that occur repeatedly (i.e. frequent patterns) [13]. HGCM utilizes FPM to calculate the prevalence of genotypes within disease populations. Genotypes are combined into patterns, and the prevalence of patterns within a predetermined population is calculated. This strategy of combining genotypes into patterns provides a more sophisticated analysis of genotype association appropriate for studying complex disease etiology, as such diseases are clearly not the product of a simple genetic event.

The algorithm eliminates patterns with such low prevalence that they cannot be relevant to the population on the whole. Though this reduces the computational load required to explore the possible patterns, an analysis of many genotypes quickly generates millions and billions of patterns to be evaluated. An arsenal of strategies is required to handle this challenge, including the intelligent reduction of data to a representative sample of genotypes and the use of Big Data infrastructure and tools to increase the computing capabilities.

### 2.2.2. *Contrast Mining*

FPM identifies patterns inherent within a population, but it provides no context for whether the patterns are expected or surprising. In other data mining use cases, the underlying data may be easily understood and insights can be gleaned simply from examining patterns. Genotypes, however, are not intuitive and therefore more context must be provided to highlight impactful results. HGCM provides this context by applying Contrast Mining.

Contrast Mining aims to identify key differences in frequent patterns between separate populations in a process involving three steps: separation, FPM, and comparison [14]. The Contrast Mining application implemented in HGCM first separates a diverse population of individuals with a complex disease into more homogeneous subgroups with shared phenotype characteristics. FPM then identifies frequent patterns in each subgroup individually, and the resulting genotype patterns are compared between the two groups. Patterns which appear frequently in one subgroup but not the other are highlighted as the key genetic distinctions between the groups.

### 2.2.3. *Inheritance Structure*

Under the hypothesis that a complex disease is caused primarily by commonly inherited genotypes, even more context can be derived by comparing the genotype patterns of disease cases to their healthy family members. Such a direct comparison between close family members begins to control for the unmeasured genetic landscape of the family and for environmental conditions. This increases the clinical relevance of the emergent genotype patterns and reinforces confidence in the results.

In the case when two healthy parents have a child with a complex disease, the influential genetic variation will most often have originated from the healthy parents. This implies that the inherited risk factors must involve interactions between other risk factors, potentially genetic, environmental, or another mechanism. Since HGCM focuses on genetic risk factors, these interactions necessarily have the form of combinations of genotypes observed in the disease patient but absent from either parent. To account for this, HGCM recalculates the prevalence of genotype patterns by omitting disease patients who have a close family member with the same pattern. This emphasizes patterns which are unique combinations of inherited genotypes, having no counterexample of a healthy family member with the same genotype pattern.

### 2.2.4. *Autism Case Studies*

The Somatic Complaints and Small Vocabulary subgroups were each contrasted against their respective outgroups, or the set of autism patients not included in the subgroup. The genetic patterns identified by the algorithm ranged from those with equal prevalence in the subgroup and outgroup to those heavily favoring one group over the other. Patterns were evaluated using the Family-Based Association Test [12] to select the ones with significant associations to the autism subgroup. These significant patterns were processed to identify the represented genes, which were then ready to be analyzed by the pathway enrichment analysis procedure.

## 2.3. Multi-Level Pathway Enrichment Analysis

Over-representation pathway enrichment analyses utilize the information within a selected set of genes to identify the pathways relevant to a research question. The output of the HGCM procedure is used as the input for pathway enrichment analyses (Figs. 2.1(a) and (b)), which ultimately selects pathways for closer examination. This section describes how over-representation pathway enrichment analyses can escalate from a basic analysis of simple gene lists to more complex implementations which utilize different strategies to glean information from lists of gene pairs.

Though these statistical methods were developed to progress research starting with the HGCM analysis, they were developed to be useful when testing a gene pair list regardless of the methodology which identified the list. Differential Gene Coexpression Analysis is another procedure which outputs lists of paired genes ideal for this multi-level pathway enrichment analysis [15].

### 2.3.1. *Single-Gene Single-Pathway Analysis (SGSP)*

The SGSP analysis is the fundamental version of the enrichment analysis, and is also the most widely used due to its availability and simplicity. Multiple statistical models can be used to perform the analysis, including the hypergeometric, binomial, chi-square, and Fisher's exact test [16]. We focus on the use of the hypergeometric distribution test to evaluate pathways for

an unexpectedly high proportion of selected genes. This test employs the
following equation:

$$P = 1 - \sum_{i=0}^{x-1} \frac{\binom{m}{i}\binom{n-m}{k-i}}{\binom{n}{k}}. \tag{2.1}$$

If $k$ genes are selected from a background set of $n$ possible genes, this
test gives the probability $P$ of observing at least $x$ of the selected genes
in a pathway with $m$ total genes. This probability is calculated for many
pathways, identifying the pathways which are implicated by the selected
genes.

Table 2.1 lists details of the parameters in Eq. (2.1), written using set
notation. This table assumes that the initial input is the same for all three
statistical tests, and therefore must be a set of gene pairs. For this basic

Table 2.1. Parameters for hypergeometric distribution test
using three levels of the pathway enrichment analysis.

| SGSP | $n$ | $|BG|$ |
|---|---|---|
| | $m$ | $|W|$ |
| | $k$ | $|\{g_1|(g_1,g_2) \in S \text{ or } (g_2,g_1) \in S\}|$ |
| | $x$ | $|\{g_1 \in W|(g_1,g_2) \in S \text{ or } (g_1,g_2) \in S\}|$ |
| GPSP | $n$ | $\frac{|BG|*(|BG|-1)}{2}$ |
| | $m$ | $\frac{|W|*(|W|-1)}{2}$ |
| | $k$ | $|S|$ |
| | $x$ | $|\{(g_1,g_2) \in S|g_1 \in W \text{ and } g_2 \in W\}|$ |
| GPPP | $n$ | $\frac{|BG|*(|BG|-1)}{2}$ |
| | $m$ | $|W_A| * |W_B|$ |
| | $k$ | $|S|$ |
| | $x$ | $|\{(g_1,g_2) \in S \mid g_1 \in W_A \text{ and } g_2 \in W_B\}|$ |

$BG$: The set of background genes.
$W$: The set of genes within a pathway, a subset of $BG$. $W_A$ and
$W_B$ are two distinct pathways.
$S$: The set of selected gene pairs $(g_1, g_2)$, where all $g_1$ and $g_2$
are in $BG$.

enrichment analysis, these are flattened into a set of individual genes for the calculation of $k$ and $x$. Note that, in practice, many analyses directly produce a set of individual genes, rather than a set of gene pairs, so for these cases this would be the only applicable version of the enrichment analysis.

### 2.3.2. *Gene-Pair Single-Pathway Analysis (GPSP)*

The basic level of the enrichment analysis can be extended to test selected pairs of genes with some biological or medical value. In the case of the GPSP analysis, selected gene pairs are used as the input instead of selected individual genes, and the enrichment analysis identifies pathways which have an unexpectedly high proportion of these gene pairs. This addresses a different question than the original enrichment analysis which is valuable for researching complex disease etiology, since such diseases are less likely to be explained by the activity of a single gene, but rather the interactions between multiple genes. Since this analysis utilizes information about gene interactions while identifying pathways, it is useful for finding pathways which must be altered multiple times to lead to disease development.

Equation (2.1) is used to perform this analysis using the variables listed in the second column of Table 2.1. In this case, $n$ is calculated to be the number of possible gene pairs, and $m$ is similarly the number of unique gene pairings in the pathway being tested. The total number of selected gene pairs is $k$, and $x$ is the number of these for which both genes in the pair are in the pathway.

### 2.3.3. *Gene-Pair Pathway-Pair Analysis (GPPP)*

Our novel extension of the enrichment analysis, the GPPP analysis, tests for the compounding effects of multiple pathways relevant to the etiology of complex diseases. In this extension, pairs of pathways are tested to evaluate the number of gene pairs connecting them. All pathway pairs are tested exhaustively, measuring the probability that the observed number of selected gene pairs which span the two pathways could have happened by chance, given the number of possible gene pairs spanning the pathways and the total number of gene pairs selected by the gene selection analysis. This analysis is designed to provide evidence that a complex disease is caused by cumulative alterations to multiple pathways.

Equation (2.1) is still used for this high-level version of the pathway enrichment analysis, using the parameters listed in the third column of Table 2.1. Similar to the GPSP level of the enrichment analysis, $n$ is the total number of possible gene pairs, and $k$ is the total number of selected gene pairs. When testing the enrichment of the connections between pathway $A$ and pathway $B$, $m$ represents the number of possible gene pairs where one gene is in pathway $A$ and the other gene is in pathway $B$. Similarly, $x$ is the number of selected gene pairs where one gene is in each of the pathways $A$ and $B$.

Note that many genes belong to multiple pathways. If these were included in the analysis, the test could unintentionally be testing the same principal as the GPSP analysis and generate unwanted results. Thus, when testing the connection between two pathways, genes with membership in both pathways are removed before identifying gene pairs which span the pathways.

### 2.3.4. *Autism Case Studies*

The results of performing the HGCM procedure on the Somatic Complaints and Small Vocabulary subgroups were used as input gene lists for the multi-level pathway enrichment analysis. Pathways enriched in the gene lists were identified by referencing the Reactome pathway database [17]. The SGSP, GPSP, and GPPP analyses were each applied to both case studies to obtain multiple perspectives for the functional pathways contributing to the etiology of the subgroups.

Figure 2.2 indicates the overlap in pathways which were significant in each level of the pathway enrichment analyses. Though there is some overlap, the Venn diagrams show that largely different sets of pathways were identified in each analysis. The strategy of analyzing a single dataset using three levels of the pathway enrichment analysis and examining the results together provide more information when prioritizing pathways for further study. It might be expected for significant pathway pairs to comprise the significant individual pathways identified by the SGSP and GPSP analyses. However, many of the GPPP results include only one pathway highlighted by another analysis. This presents a unique opportunity to investigate the

**Somatic Complaints Subgroup**     **Small Vocabulary Subgroup**

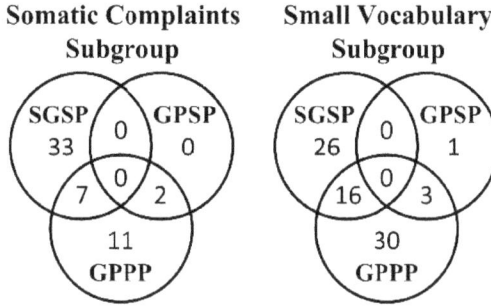

Fig. 2.2. Venn diagrams showing the overlap in the pathways highlighted by performing multiple levels of pathway enrichment analysis on the Somatic Complaints Subgroup and the Small Vocabulary Subgroup.
*Note*: SGSP = Single-Gene Single-Pathway analysis, GPSP = Gene-Pair Single-Pathway analysis, GPPP = Gene-Pair Pathway-Pair analysis.

compounding effects of multiple pathways for their connection to disease etiology.

## 2.4. Pathway Prioritization

Analyzing a dataset with all three of these described pathway enrichment analyses presents the unique opportunity to combine multiple results to prioritize pathways for further investigation (Fig. 2.1(c)). Pathways highlighted by more than one enrichment analysis are particularly interesting, as this indicates that they are likely influential on their own, but also that they have important cumulative effects along with other pathway disruptions. The strategy for merging the results of the three pathway enrichment analysis levels to prioritize pathways involves the calculation of an initial pathway priority which is subsequently modified into a final pathway priority by incorporating the initial priority score of connected pathways.

### 2.4.1. *Initial Pathway Priority*

The following formula calculates the initial pathway priority based on the results of the three pathway analysis levels.

$$S_W = -\log\left(\operatorname*{argmin}_{x \in \{\text{SGSP, GPSP, GPPP}\}} P_{W,x}\right) + \alpha * N_W + \beta * M_W, \quad (2.2)$$

where $S_W$ is the score for the pathway $W$, $P_{W,x}$ is the $P$-value for pathway $W$ using enrichment analysis $x$, $N_W \in [0, 3]$ is the number of enrichment analysis levels producing a significant $P$-value for the pathway $W$, $M_W$ is the number of other pathways paired with $W$ within the significant GPPP results, and $\alpha$ and $\beta$ are constant weights. This formula first chooses the most significant $P$-value for this pathway from the three enrichment analyses and performs the negative log to express the value as a positive score which increases with greater significance. The priority for $W$ is increased when it appears in the results of multiple enrichment analyses ($N_W$) and when the GPPP analysis connected it to multiple pathways other than $W$ ($M_W$). These factors are respectively weighted by $\alpha$ and $\beta$, to adjust the prioritization calculation according to research goals.

### 2.4.2. *Final Pathway Priority*

The initial pathway priority is modified to calculate the final pathway priority by accounting for significant GPPP pairs that indicate an interaction with other highly prioritized pathways. If $C$ is the set of all pathways connected to $W$ using the GPPP analysis, the maximum score $S_x$ for $x \in C$ contributes to the final score for $W$ using the following equation:

$$S'_W = S_W + \gamma * \operatorname*{argmax}_{x \in C} S_x, \tag{2.3}$$

where $S'_W$ is the final priority score for pathway $W$. This factor is weighted by $\gamma$ to adjust the impact of this final scoring modifier.

This formula combines the results from the triad of enrichment analyses into one cohesive prioritized list of pathways for future examination. In our applications below, we set the weights to $\alpha = \beta = \gamma = 0.25$ to balance the contributions of the three priority score modifiers.

### 2.4.3. *Autism Case Studies*

Pathway priorities were calculated for the Somatic Complaints and Small Vocabulary subgroups, ranking the results of the multi-level pathway enrichment analysis to guide future study. Table 2.2 lists these ranked results, including the priority score calculated for the top Reactome pathways and

Table 2.2.   Prioritized pathways relevant to autism subgroups.

| | **Reactome ID** | **Reactome pathway** | **Priority score** | **Related functionality** |
|---|---|---|---|---|
| Somatic complaints subgroup | R-HSA-382551 | Transport of small molecules | 7.41 | Synaptic function |
| | R-HSA-170822 | Regulation of glucokinase | 5.94 | Metabolism[a] |
| | R-HSA-2990846 | SUMOylation | 5.05 | Synaptic function |
| | R-HSA-3108232 | SUMO E3 ligase activity | 5.05 | Synaptic function |
| | R-HSA-3781865 | Diseases of glycosylation | 5.05 | Neurological activity |
| | R-HSA-1912420 | Pre-NOTCH processing in Golgi | 5.01 | Synaptic function |
| | R-HSA-1912422 | Pre-NOTCH expression and processing | 4.55 | Synaptic function |
| | R-HSA-157118 | Signaling by NOTCH | 4.30 | Synaptic function |
| Small vocabulary subgroup | R-HSA-2173793 | Transcriptional activity of SMAD | 9.76 | Cell junction TGF-beta pathway |
| | R-HSA-170834 | Signaling by TGF-beta receptor complex | 8.39 | Cell junction TGF-beta pathway |
| | R-HSA-202403 | TCR signaling | 7.33 | Immune system[a] |
| | R-HSA-8866652 | Synthesis of active ubiquitin | 6.03 | Protein degradation |
| | R-HSA-166662 | Lectin pathway of complement activation | 5.50 | Immune system[a] |
| | R-HSA-186712 | Regulation of beta-cell development | 5.50 | Metabolism[a] |
| | R-HSA-166658 | Complement cascade | 5.25 | Immune system[a] |
| | R-HSA-166663 | Initial triggering of complement | 5.25 | Immune system[a] |

*Note:*[a]Pathways not closely associated with autism according to Ref. [18].

the general functional classification of the pathway. These pathways were cross-referenced with a recent review of autism genetics and pathways to assess whether they are known to be relevant to autism development [18]. Highlighted pathways with no close connection to those discussed in the review are noted in the table.

Small Ubiquitin-like Modifier (SUMO) post-translational modification is closely tied to synapse formation [19] and the Notch cell surface receptor is involved in the regulation of nervous system connectivity [20]. Therefore, all but one of the pathways prioritized for their relevance to the "Somatic Complaints" subgroup can be categorized into known autism functionality, particularly with synaptic function. The results of the GPPP analysis supplement this knowledge of autism mechanisms by implicating the highest prioritized pathway, transport of small molecules, in conjunction with Notch processing. This connection should be investigated as the specific mechanism by which Notch signaling produces autistic phenotypes, particularly phenotypes relating to somatic complications.

The results relating to the "Small Vocabulary" subgroup indicate much different pathways relevant to this group of autism patients. The cell junction TGF-beta pathway and protein degradation are likely relevant to autism due to their roles in synapse regulation and activity [21], which parallels the results seen for the Somatic Complaints subgroup. The other results distinguish the two groups, as several of the top prioritized pathways for this subgroup in Table 2.2 relate to the immune system, a function which is not listed as a main biological pathway associated with autism according to [18]. However, other reviews discuss a strong connection between autism cohorts and immune system deficiencies, largely relating to anti-brain antibodies [22, 23]. Our results suggest that these immune-related mechanisms may be especially impactful for autism subgroups with limited vocabularies and other language development deficiencies.

## 2.5. Discussion

Transforming biological data into medical knowledge is a hurdle that must be overcome to promote progress. AI technology connecting gene-level data to pathway-level information can assist with this transition, but biomedical applications require that the process be explainable from start to finish to instill confidence in the results. Frequent Pattern Mining and Contrast Mining introduce the computational power of AI while retaining explainable methodology and results. These algorithms are supplemented by sound statistical tests to further increase the reliability of results, starting with the Family-Based Association Test to distinguish genetic patterns which distinguish populations and moving forward to the multi-level pathway enrichment analysis which also utilizes robust but understandable statistical practices. None of the methods discussed here are "black-box", which is critical to allow the results to be usable in future stages of research.

HGCM is a data-driven method for detecting associations between groups of diseased individuals and underlying genetic patterns. It can be utilized to reveal the genetic mechanisms driving the development of a complex disease, and it excels at performing focused examinations of the differences between patients grouped by shared characteristics. However, the identification of the genes relevant to disease development is only the beginning of the path to understanding disease mechanisms and ultimately developing technologies to assist in detection, prevention, and treatment. To drive research of complex disease etiology, the next step is to identify functional mechanisms implicated by disease-associated genetic patterns.

There is a research bottleneck when deciding how to partition the raw data before analysis even begins. Identifying criteria for dividing a complex disease population into promising subgroups is still a manual process that relies on the qualitative assessment of clinical experts to make that determination. The introduction of an exploratory analysis that can examine many disease subgroups and provide guidance for which subgroups should be analyzed in more depth would speed up these early stages of research. Furthermore, the subgroups identified for deeper analysis would already be supported by preliminary data-driven insight that they are meaningful at a genetic level.

The algorithms in the current analysis pipeline are a useful initial integration of AI into bioinformatics procedures, but they need to be improved and expanded to support the next generation of biomedical research. The genotype patterns identified by the Contrast Mining algorithm to distinguish groups of patients are not fully utilized by the subsequent analyses.

The multi-level pathway enrichment analysis presented here begins to leverage this advantage by testing lists of gene pairs, but larger combinations of genotypes remain an untapped resource. Much of the challenge is the difficulty interpreting larger sets of connected genes, but this presents an opportunity for future AI algorithms to interpret patterns into information which is likely to be adopted by experts during their decision making process.

Finally, there is room for new AI implementations to extend this analysis pipeline, pushing the automated output even closer to actionable results. There are likely multiple directions these extensions could go while utilizing the prioritized pathways which currently mark the end of the analysis pipeline. One promising direction is the automatic analysis of pathways for the purpose of drug repositioning. Pathways determined to be relevant to a complex disease could be queried in a database of drug targets. This would identify medications with a potential for application as a treatment for new diseases, despite their original intended target.

This chapter demonstrates how AI has already advanced the genotype analysis pipeline methodology to allowing the investigation of genetic contributions to heritable diseases, but there are still many aspects of the process that can benefit from automation and intelligent computation. Areas for further AI involvement span from the very beginning of the analysis pipeline to the end, and could even be used to extend the scope of automated processes. Ultimately, AI should be utilized to the greatest extent possible to increase the velocity of biological knowledge and the development of medicine.

## References

1. Schmitt, M. W., Kennedy, S. R., Salk, J. J., Fox, E. J., Hiatt, J. B. and Loeb, L. A. (2012). Detection of ultra-rare mutations by next-generation sequencing. *Proceedings of the National Academy of Sciences* 109, pp. 14508–14513.
2. Glazko, G. V. and Emmert-Streib, F. (2009). Unite and conquer: Univariate and multivariate approaches for finding differentially expressed gene sets. *Bioinformatics* 25, pp. 2348–2354.
3. Ivanov, H. Y., Stoyanova, V. K., Popov, N. T., and Vachev, T. I. (2015). Autism spectrum disorder: A complex genetic disorder. *Folia Medica* 57, pp. 19–28.
4. Hornberg, J. J., Bruggeman, F. J., Westerhoff, H. V. and Lankelma, J. (2006). Cancer: A systems biology disease. *Biosystems* 83, pp. 81–90.
5. Flannick, J. and Florez, J. C. (2016). Type 2 diabetes: Genetic data sharing to advance complex disease research. *Nature Reviews Genetics* 17, p. 535.

6.   Liu, Z.-P., Wang, Y., Zhang, X.-S. and Chen, L.-N. (2012). Network-based analysis of complex diseases. *IET Systems Biology* 6, pp. 22–33.
7.   Billings, L. K. and Florez, J. C. (2010). The genetics of type 2 diabetes: What have we learned from GWAS? *Annals of the New York. Academy of Sciences* 1212, pp. 59–77.
8.   Gaugler, T., Klei, L., Sanders, S. J., Bodea, C. A., Goldberg, A. P., Lee, A. B., Mahajan, M., Manaa, D., Pawitan, Y. and Reichert, J. (2014). Most genetic risk for autism resides with common variation. *Nature Genetics* 46, pp. 881–885.
9.   Miles, J. H. (2011). Autism spectrum disorders — a genetics review. *Genetics in Medicine* 13, p. 278.
10.  Fischbach, G. D. and Lord, C. (2010). The Simons Simplex Collection: A resource for identification of autism genetic risk factors. *Neuron* 68, pp. 192–195.
11.  Achenbach, T. M. (1994). Child Behavior Checklist and related instruments. In M. E. Maruish (ed.), *The Use of Psychological Testing for Treatment Planning and Outcome Assessment*, Hillsdale, NJ, US: Lawrence Erlbaum Associates, Inc, pp. 517–549.
12.  Spencer, M., Takahashi, N., Chakraborty, S., Miles, J. and Shyu, C.-R. (2018). Heritable genotype contrast mining reveals novel gene associations specific to autism subgroups. *Journal of Biomedical Informatics* 77, pp. 50–61.
13.  Hipp, J., Güntzer, U. and Nakhaeizadeh, G. (2000). Algorithms for association rule mining — a general survey and comparison. *ACM Sigkdd Explorations Newsletter* 2, pp. 58–64.
14.  Bay, S. D. and Pazzani, M. J. (2001). Detecting group differences: Mining contrast sets. *Data Mining and Knowledge Discovery* 5, pp. 213–246.
15.  McKenzie, A. T., Katsyv, I., Song, W.-M., Wang, M. and Zhang, B. (2016). DGCA: A comprehensive R package for differential gene correlation analysis. *BMC Systems Biology* 10, p. 106.
16.  Khatri, P. and Drăghici, S. (2005). Ontological analysis of gene expression data: Current tools, limitations, and open problems. *Bioinformatics* 21, pp. 3587–3595.
17.  Fabregat, A., Korninger, F., Viteri, G., Sidiropoulos, K., Marin-Garcia, P., Ping, P., Wu, G., Stein, L., D'Eustachio, P. and Hermjakob, H. (2018). Reactome graph database: Efficient access to complex pathway data. *PLoS Computational Biology* 14, p. e1005968.
18.  Huguet, G., Benabou, M. and Bourgeron, T. (2016). The genetics of autism spectrum disorders. In: *A Time for Metabolism and Hormones*. Springer, pp. 101–129.
19.  Schorova, L. and Martin, S. (2016). Sumoylation in synaptic function and dysfunction. *Frontiers in Synaptic Neuroscience* 8, p. 9.
20.  Giniger, E. (2012). Notch signaling and neural connectivity. *Current Opinion in Genetics & Development* 22, pp. 339–346.
21.  De Rubeis, S., He, X., Goldberg, A. P., Poultney, C. S., Samocha, K., Cicek, A. E., Kou, Y., Liu, L., Fromer, M. and Walker, S. (2014). Synaptic, transcriptional and chromatin genes disrupted in autism. *Nature* 515, p. 209.
22.  Masi, A., Glozier, N., Dale, R. and Guastella, A. J. (2017). The immune system, cytokines, and biomarkers in autism spectrum disorder. *Neuroscience Bulletin* 33, pp. 194–204.

23.  Meltzer, A. and Van de Water, J. (2017). The role of the immune system in autism spectrum disorder. *Neuropsychopharmacology* 42, p. 284.
24.  Visscher, P. M., Brown, M. A., McCarthy, M. I. and Yang, J. (2012). Five years of GWAS discovery. *American Journal of Human Genetics* 90, pp. 7–24.

Chapter 3

# Blockchain for Pre-clinical and Clinical Platform with Big Data

Yin-Wu Chen and Zon-Yin Shae

*Department of Computer Science and Information Engineering,
Asia University, Taiwan*
*Pervasive Artificial Intelligence Research (PAIR) Labs, Taiwan*

## Abstract

Big Data analytics requires consistent data interpretation and unified data format in order to apply data computation algorithms to extract useful information, detect/predict trending and make improve the decision-making process. Health care has been one of the many fields medical or medicine data are collected in very divided format and, in many occasions, with subjective data interpretations [1, 2].

The research work looked into a newly developed system, the Integrated Biomedical Informatics System (IBIS/BRICS) [3] funded and developed by the National Institutes of Health (NIH), to understand how the data collection process can be improved with a creation of common data dictionary (CDE) and user definable form structure. We, then, proposed a software architecture to use Ethereum Blockchain [4] network to manage these collected datasets such that various organizations can participate in the network, and securely as well as reliably share medical results with each other. A huge medical data platform can therefore be created for big data analytics. We imported Taiwan Stroke Registry [5] data to IBIS/BRICS and exported them via a Blockchain Adapter onto Ethereum Blockchain network for management and future Big Data Analysis.

## 3.1. Introduction

We use IBIS/BRICS system with standard common data dictionary (CDE) and user definable form structure for data collection as our first experimental data front-end, explore the benefits of dataflow built into IBIS/BRICS, and bring the IBIS/BRICS dataset to an Ethereum Blockchain platform deployed with interworking smart contracts. These smart contracts can manage the dataset access by various organizations participating on the network. New smart contracts can be developed in the future for the Big Data analytics.

Figure 3.1 demonstrates a simplified software architecture diagram for a Blockchain network such as Ethereum Blockchain to interface with IBIS/BRICS. In this figure, we specified a Blockchain Adapter to extract dataset from IBIS/BRICS, encrypt it, hash it, and post dataset management information (e.g. encryption key, hash string and document ID, etc.) onto Blockchain. Any user having access right to the Blockchain and the dataset will be able to retrieve the dataset, and automatically verify its genuineness before putting to use. Each legacy informatics system or each organization will be designed with a Block Adapter to bring the dataset onto a common Blockchain network for management and sharing. Thus, a single Blockchain Adapter failure or compromise disables only transactions in and out of the owning organization. All other transactions among other organizations can still continue without interruption. This is the advantage of a de-centralized transaction provided by Ethereum Blockchain.

In the future, Fig. 3.1 architecture will evolve into a Blockchain-centric architecture as shown in Fig. 3.2 where the external legacy system is eliminated for better security and reliability and is replaced with a set of collaborative smart contracts interworking to provide equivalent functions provided by IBIS/BRICS and Blockchain Adapter of Fig. 3.1.

In practice, a working architecture shown in Fig. 3.3 is being developed. We used the Integrated Biomedical Informatics System (IBIS) to create data dictionary and form structures to import Taiwan Stroke Registry data. An intermediate system, Mirth Connect (https://en.wikipedia.org/wiki/Mirth_Connect) [6, 7] is applied to filter and transform the data from IBIS/BRICS and output to a new Blockchain Adapter, where dataset feature extraction, cryptography, and hashing are performed and posted onto Blockchain.

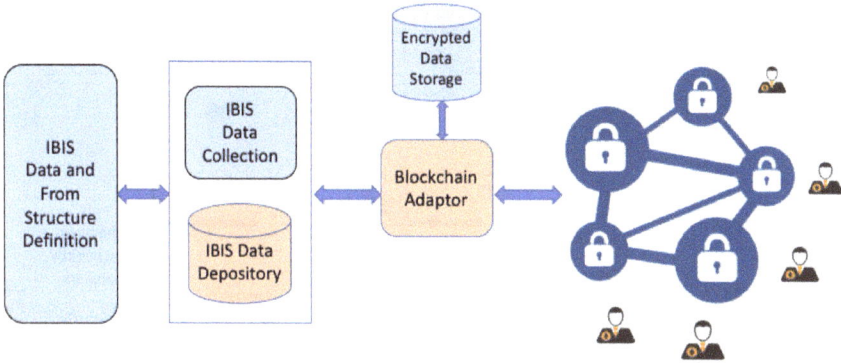

Fig. 3.1. A simplified block diagram for **IBIS/BRICS** and Ethereum Blockchain Integration.

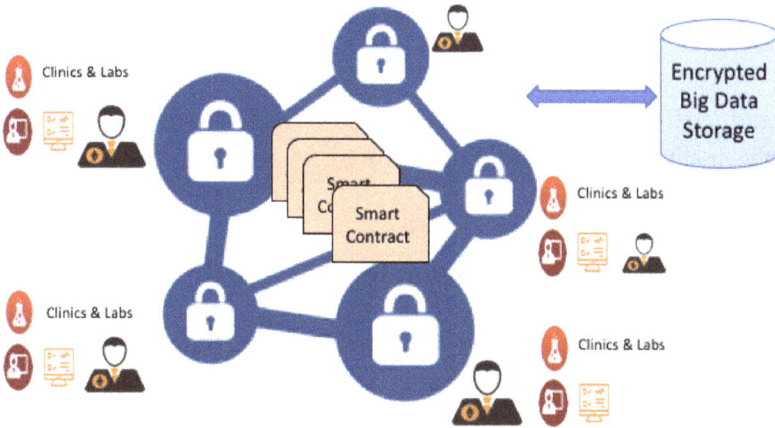

Fig. 3.2. Ethereum Blockchain-Centric Informatics System.

The following sections describe the dataflow and functions of IBIS/BRICS and Mirth Connect and specify the software components and data logic required for Blockchain Adapter. The benefits of data dictionary and form structure of IBIS/BRICS are identified. The versatility of Mirth Connect is exploited. The smart contracts deployed in the Ethereum Blockchain are described in later sections.

Though the architecture for this research is focused on the management of Stroke Registry data or, in more general terms, the Electronic Healthcare

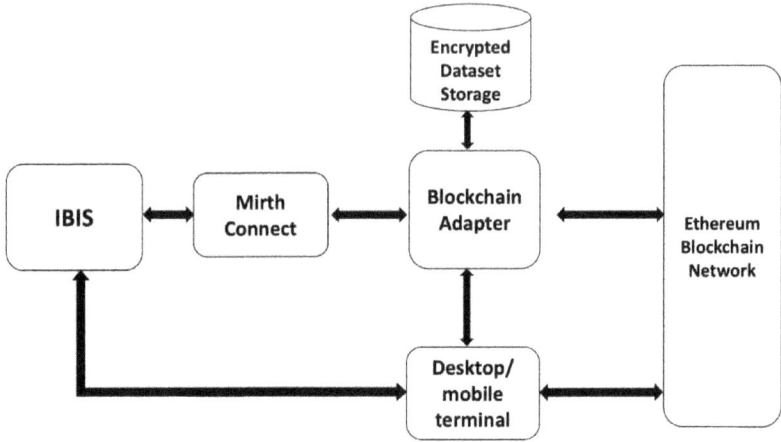

Fig. 3.3.   A working architecture for IBIS/BRICS and Ethereum Blockchain Integration.

Record (EHR) data, the architecture can be easily generalized for dataset management in other fields of application.

## 3.2.   Integrated Biomedical Informatics System (IBIS)

The Integrated Biomedical Informatics System (IBIS) is a collaborative project funded by multiple institutes at the National Institutes of Health (NIH) along with the Center for Neuroscience and Regenerative Medicine (CNRM) and Department of Defense (DOD) to build a reusable and sustainable informatics infrastructure [8] to promote and catalyze biomedical translational research and clinical trial. Instead of focusing on the application and software tools in the conventional system, the system focuses on having the common data dictionary and user definable data capture instruments and tools to achieve sharing diverse and heterogeneous data, and clinical phenotype data through research networks. The system is customizable for different organizations, governmental or non-governmental, with aims for better software reusability and sustainable software team. The model-view-controller (MVC) design pattern [9] is used for developing the IBIS/BRICS software to allow for efficient code reuse and parallel development.

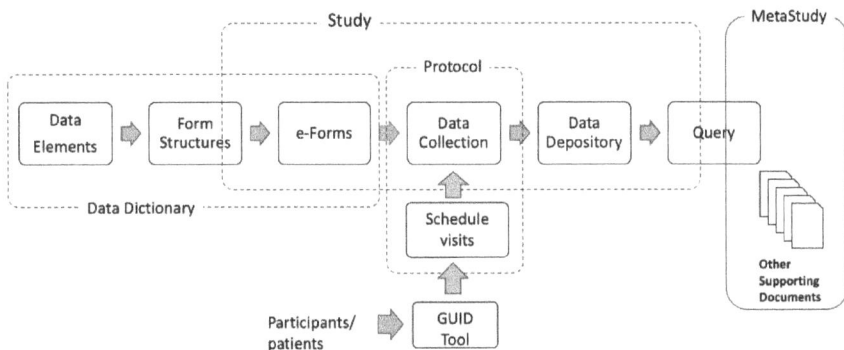

Fig. 3.4.   IBIS/BRICS dataflow.

### 3.2.1.   *IBIS/BRICS Dataflow*

Figure 3.4 shows the data flow IBIS/BRICS user needs to follow to create a study case, capture data, and share study results.

When researchers define a clinical or pre-clinical study, they usually come up with a set of information they want to collect from a subject. IBIS/BRICS dictionary module provides tools for defining such elements for data collection. Researcher can use the dictionary module to create data elements and organizes them into one or multiple form structures followed by publishing them as electronic forms (e-Form). Data Dictionary plays a very important role in the IBIS/BRICS design. Data elements defined as Common Data Element (CDE) are commonly used by the clinical and pre-clinical community and have common interpretations. Since 2010, NINDS (National Institute of Neurological Disorders and Stroke) CDE Workgroup has developed a set of fundamental CDE to serve all disease areas. Each organization participating in IBIS/BRICS program can incorporate the NINDS set and extend it using aliasing and translation rule for same elements. Though new data elements can also be defined to capture special data not defined in the CDE, to broaden the utility of the TBI CDEs, experts were asked to update the recommendations to make them relevant to all ages, injury severity, and phases of recovery. The CDE developed and maintained by NINDS can be found at the following link: https://www. commondataelements.ninds.nih.gov/TBI.aspx#tab=Data_Standards.

### 3.2.2.  *Data Collection in IBIS*

Researchers begin collection by creating a protocol for the study, and then add subjects and schedule visits for collecting data. After the scheduled data collections are completed, the collected data can be validated against the data dictionary and submitted to Data Repository of IBIS/BRICS using a Submission Tool. Other researchers can then access the shared study results from Data Repository using the Download Tool. On a programmable schedule, data in Data Depository will be downloaded to a graph database (Virtuoso) for permanent storage. The Query Tool querying the Virtuoso database can be used to assemble multiple study results into a so-called query-able package. User can then create a Meta Study to bundle these query-able packages and other supporting documents for additional analysis or future publications. Inside the Query Tool, user can also browse and filter the dataset to check on the results.

### 3.2.3.  *System Modules of IBIS*

Current IBIS/BRICS has six major visible software application modules arranged on graphics interface as shown in Fig. 3.5.

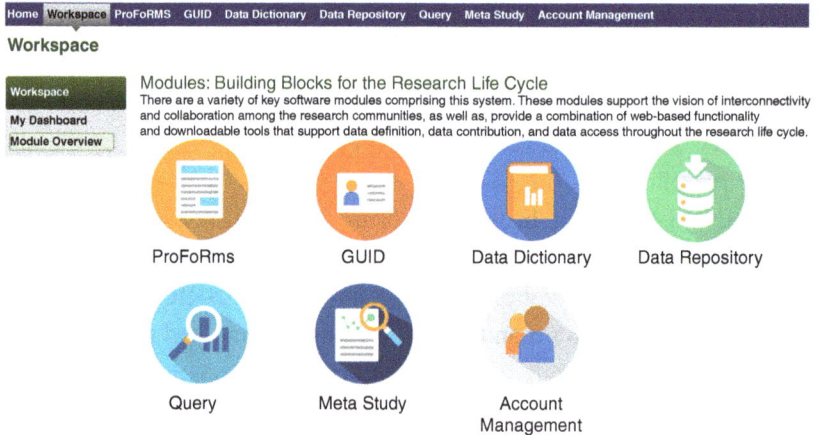

Fig. 3.5.   Main IBIS/BRICS menu screen.

**Data Dictionary** defines data elements, form structures, and e-forms. **ProFoRm** manages protocols, subjects, and visits for collecting data also report and query results. **Data Repository** manages study to include form structures, and provide tools for dataset validation, upload (submission) and download. **Query Tool and Meta Study** manage multiple datasets into query-able datasets and incorporate them into meta study for additional analysis and publication reference. **Account Management manages** user accounts, privileges, and access to different module or datasets. **GUID (Global User ID)** generate GUID or pseudo GUID from subject demographics data for the system use, which protects personal identity information from revealing to unrelated users.

Behind the scene, IBIS/BRICS uses PostgreSQL database for Data Dictionary, ProFoRMS, and Data Repository applications. Virtuoso graph database is used as a dataset vault from which query tool can assemble data collected from multiple studies into a data cart for viewing and sharing.

From the software development point of view, IBIS/BRICS applies the MVC (model-view-controller) pattern [9] in the module design. Most of the codes were developed in Javascript, JQuery, and Java language. The user authentication is based on Spring Security [10] interaction between a Central Authentication Server (CAS) and all software applications for single sign-on, i.e. signing to the system on one application will automatically authenticate access to other applications. The https scheme is supported for all web services among system modules. Account management application provides nine individual privileges assignable to every account user. The main menu items will be dimmed for no function when the associated privileges are not available.

Figure 3.6 shows the server architecture for different software modules. Depending on the database size and server configuration, PostgreSQL databases can be consolidated into one or multiple database instances in one or few virtual or physical server instances, same for Virtuoso database and all software application modules.

It is worth mentioning that a special module in IBIS/BRICS is used for protecting the privacy of study participants. Global User ID (GUID) tool

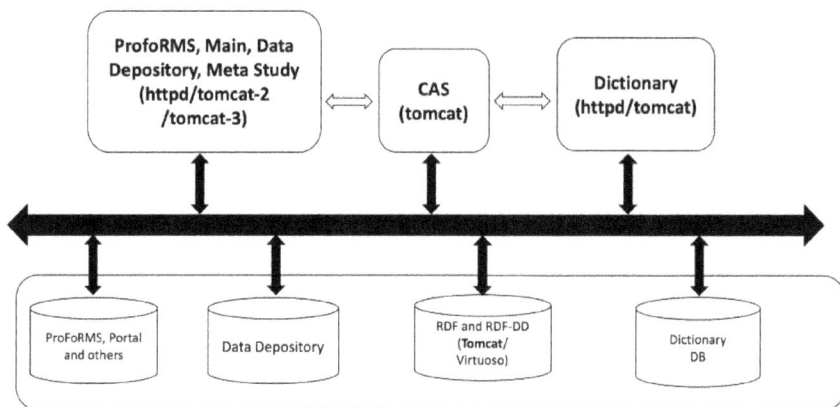

Fig. 3.6.   IBIS/BRICS server architecture.

can generate a GUID for a subject based on a set of Personal Identification Information (PII). The current PII contains name, date-of-birth, place-of-birth, gender, and national ID. However, if, for some reasons, the PII is not completely available for generating a valid GUID, a pseudo GUID can be generated for use instead. There is a safe-proof mechanism in the GUID tool. When the PII nearly matches with the other PIIs in the system, user will be prompted for the minor discrepancy and proposed to either double check to correct PII error, or simply go ahead with the GUID generation using the near-matched PII. This process will avoid orphanage data due to PII typo when using GUID tool.

IBIS/BRICS based data management system is to support new and existing research and resource development promoting biomarker discovery. It has been deployed for the Parkinson's Disease Biomarkers Program (PDBP) at NINDS (https://pdbp.ninds.nih.gov), Federal Interagency Traumatic Brain Injury Research (FITBIR) (https://fitbir.nih.gov), and The National Ophthalmic Disease Genotyping and Phenotyping Network (https://eyegene.nih.gov), etc. Our research participation on exploring IBIS/BRICS installation in Taiwan is to use it as a data depository for Taiwan Stroke Registry data. The end goal is to build an Ethereum Blockchain network to provide a de-centralized, more secure, more reliable, and trustful data exchange for Big Data analytics on Stroke Data.

## 3.3. Mirth Connect

As described in the user manual [10] of Mirth Connect, and as shown in Fig. 3.7:

> "Mirth Connect open source software is an integration engine that can receive data from a variety of sources and take powerful actions on that data, including sending the data out to multiple external systems. It can also transform data from one format to another, or extract pieces of the data you can act on or send outbound."

Mirth Connect builds on the concept of data channel by which data are received by a variety of methods, filtered, transformed, and sent out to multiple destinations. The series of programmable receiver, filtering and transformation forms a flexible architecture for data verification, cleansing, translation, transformation, and feature extractions. These are all popular functions we ever need for Big Data analytics. Mirth Connect architecture provides the capability for us to add customized functions in the data pipeline. Here is a list of data inbound and data outbound mechanisms supported by Mirth Connect.

For inbound data: DICOM listener, Database Reader, File Reader, HTTP Listener, JMS Listener, JavaScript Reader, TCP Listener, and Web Service Listener.

For data outbound: Channel Writer, DICOM Sender, Database Writer, Document Writer, File Writer, HTTP Sender, JMS Sender, JavaScript Writer, SMTP Sender, TCP Sender, and Web Service Sender.

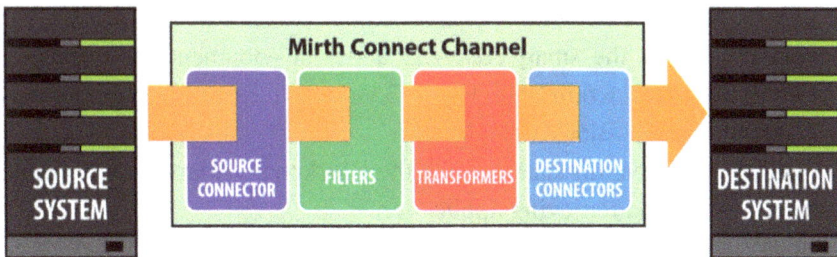

Fig. 3.7. Mirth connect high level dataflow diagram.

Additional support for Email Reader, Serial Connector, and Mirth Results Connector, etc., are available as commercial packages.

In this research project with IBIS, Mirth Connect channel will be configured to read exported files from IBIS/BRICS data repository. These files contain the existing Taiwan Stroke Registry data imported into IBIS/BRICS data repository. Mirth Connect channel will output dataset as XML messages to Blockchain Adapter for metadata extraction, dataset encryption, hashing and posting onto Ethereum Blockchain. The architecture calls for one Mirth Connect server to interface each informatics system such as IBIS. One Blockchain Adapter is paired with one Mirth Connect to bring datasets into Ethereum Blockchain network.

The role of Mirth Connect in this project can be listed as follows:

1. Upon triggering by IBIS/BRICS or user, execute one or multiple channels to read datasets from IBIS. These datasets spread over multiple files in csv or xml format.
2. Verify and cleanse the data elements based on a set of verification and data cleaning algorithm.
3. Filter dataset to extract dataset ID, time stamp of data creation, owner ID, and subject GUID.
4. Transform dataset to a pre-defined xml format.
5. Send dataset to Blockchain Adapter along with dataset ID, time stamp of data creation, owner, and subject GUID.

### 3.4. Blockchain Adapter and Smart Contracts in Ethereum Blockchain Network

Figure 3.8 shows the smart contracts under development for managing stroke dataset collected in IBIS. Figure 3.9 shows the software components implemented in the Blockchain Adapter.

Here is the operating data logic in the Blockchain Adapter for a simple dataset posting and retrieving application:

1. Upon receiving a dataset from Mirth Connect, Blockchain Adapter will request a key from Data Manager contract for the subject GUID.

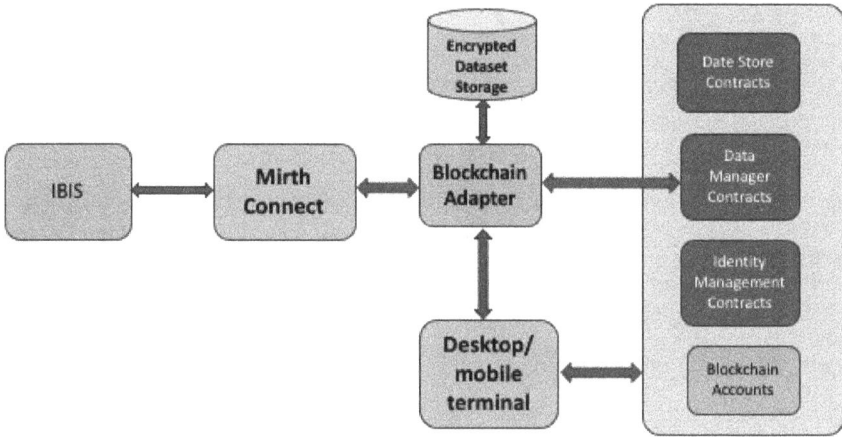

Fig. 3.8.   Smart contracts for stroke dataset management.

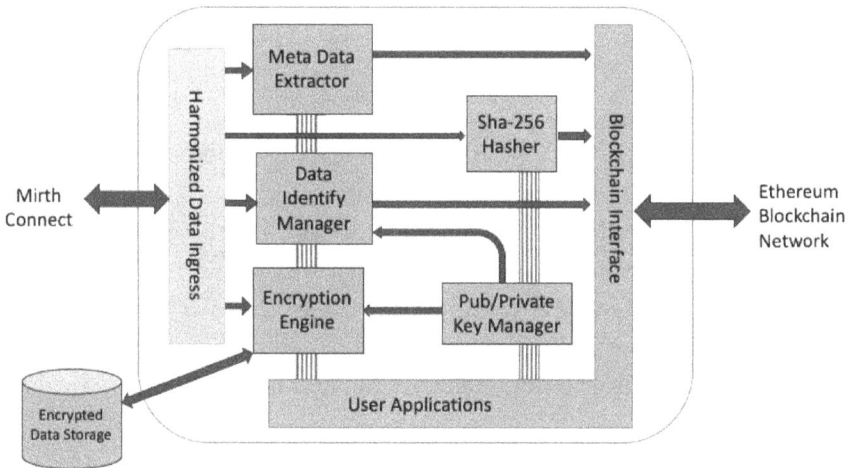

Fig. 3.9.   Functional blocks of block adapter.

2. Encrypt incoming dataset for the subject GUID and store the encrypted dataset to the Encrypted Data Storage server.
3. Hash the encrypted dataset to obtain a hash string called Dataset Hash.
4. Post subject GUID, dataset ID, time stamp of data creation, dataset owner, and Dataset Hash to Ethereum Blockchain via designated Data

Manager smart contract. Data shall be stored in a Data Store smart contract for the subject GUID.

5. Upon requesting for a dataset from a data requester by subject GUID and dataset ID, check with Data Manager smart contract for the requester's right to access the dataset.
6. If the requester has the access right, make a transaction with Data Manager smart contract to retrieve dataset hash and decryption key for the subject GUID.
7. Obtain the encrypted dataset by the dataset ID and subject GUID, decrypt it and present it to the data requestor. Erase key after decrypting the dataset.
8. Data Manager contract manages the dataset access right only on a per subject GUID basis.

From the security and reliability point of view, a few design guidelines are strictly followed for the Blockchain Adapter and smart contracts deployed to Ethereum Blockchain:

1. Never store private key and public key in the same database or into the same smart contract, nor transport them together in the same communication channel.
2. Private key stored in Data Store contract per subject GUID can only be accessed by the Data Manager contract which creates it. Private key security can be further enforced by encryption with a key privately held by Data Manager smart contract.
3. Data access right shall be managed by the Data Store per individual subject GUID.
4. Use Identity Manager smart contract as a backup mapping between subject GUID and its private key.
5. Blockchain Adapter shall not store any information regarding public/ private keys, Dataset Hash, smart contract addresses, and Blockchain account password. Upon startup, Blockchain Adapter shall prompt user for the password of a preset blockchain account. All transactions made by Blockchain Adapter with Blockchain shall use this account. This will ensure a clean startup in case of Blockchain Adapter failure or hacker attacks. When one Blockchain Adapter fails, transactions (dataset

exchange) with the associated organization will be interrupted, but other transactions among all other organizations shall not be affected.

Blockchain Adapter in this research project is implemented in a node.js server. The web3.js module is used to interface with Ethereum Blockchain. Blockchain Adapter also serves web services for desktop and mobile apps.

## 3.5. Smart Contracts State and Function Details

For the simple dataset posting and retrieving application, smart contracts in this architecture will store state data and provide functions as follows.

### 3.5.1. *Data Manager Smart Contract*

Each legacy informatics will be assigned a Data Manager Contract to inter-act with Ethereum Blockchain. Data Manager contract will store subject GUID, Data Store contract address for the subject GUID. When posting dataset to Ethereum Blockchain, Data Manager will supply the public key to the Blockchain Adapter for encryption if the subject GUID already exists with a valid private key in the associated Data Store contract. Otherwise, Data Manager will produce a private/public key pair, give the public key to Blockchain Adapter for encryption, create a Data Store contract for the subject GUID and save the private key in the Data Store contract.

In the process of retrieving dataset, Data Manager will look up for the subject GUID and Data Store contract address, check if the requesting Blockchain account is in the authorized list for the subject GUID. If it is, the Data Store will return a suitable key, available dataset ID and time stamp to the Data Manager, then Blockchain Adapter. If not, appropriate message will be returned to Data Manager, then Blockchain Adapter.

### 3.5.2. *Data Store Smart Contract*

Data Store contract will store subject GUID, authorized keys for the subject GUID, authorized account addresses, dataset ID, time stamp of dataset creation, and some metadata for the dataset. There will be multiple dataset IDs in each of the Data Store contract. That means a single subject GUID can have multiple records of healthcare dataset collected at different time in different hospitals. And, they will all be encrypted with a public key which can be derived from the authorized keys store in the Data Store contract.

### 3.5.3. *Identity Management Smart Contract*

Identity management contract will store Blockchain Adapter ID (representing a single organization or a legacy informatics system in the integration architecture), Blockchain account address for the Blockchain Adapter, Data Manager contract address, subject GUID, and private key under management. This smart contract is intended for bookkeeping the information of smart contracts in support of a single Blockchain Adapter. The design scenario for setting up a Blockchain Adapter into operational state is, first, create an Ethereum Blockchain account for the Blockchain Adapter; second, deploy an Identity Management smart contract with a pre-defined Blockchain Adapter ID and the newly created Blockchain account address. The third, call an initialization function in the Identity Management contract to deploy a Data Manager contract for Blockchain Adapter to work with.

Given this smart contract architecture, there will be one Blockchain Adapter contract, one Data Manager contract, one Identity Management contract deployed for every Blockchain Adapter in the production environment. There will be millions of Data Store contracts for all participating subject GUIDs plus additional GUIDs resulting from the discrepancy of personal data used to generate the GUID. Performance and scaling of the smart contract are not covered in this research. Before production deployment, overall performance such as transaction latency, smart contract size etc., should be carefully assessed and benchmarked to the pair up with latest development on Ethereum Blockchain technology in the community.

### 3.6. Blockchain Network-centric Architecture

Medical informatics system such as IBIS/BRICS uses instances of databases in the cloud or non-cloud environment to store and provide data exchange is considered operating in a centralized application environment. That means the overall architecture shown in Fig. 3.8 is still subject to IBIS/BRICS failures in space (across servers and sites) and in time (scheduled time of data capture).

Blockchain network such as Ethereum has been recognized as the next generation Decentralized Application (DAPP) platform for a more secure, more transparent, and more reliable information exchange. It can deliver such promises based on the following few fundamental principles:

1. The Blockchain network consists of many homogenous nodes storing a chain of same data blocks to record all transactions (information exchanges if you will) occurring in the network. Therefore, a compromised node (failure or being hacked) will not stop new transactions from occurring in the network.
2. Success of transaction is based on a consensus algorithm running at each node and collaborated among all nodes. Failure of consensus will reverse the transaction as was never attempted before. This will guarantee that transactions will never be dictated by a single or few nodes that are compromised or hacked. Receivers of the transaction can believe that the information being exchanged is a true version.
3. Since encryption will be used to protect data being transacted, and the chain of data block is self-auditable for unwanted tempering, no small piece of data in any block, once committed, can be modified. Thus, all transactions are secured and traceable.
4. Availability of smart contracts to provide data access management in Ethereum Blockchain opens up tremendous opportunities for distributed applications that are secure, reliable, transparent, and trustful.

As Ethereum Blockchain continues to be improved, a Blockchain-centric architecture as shown in Fig. 3.10 would be the ideal future architecture for optimal healthcare informatics.

In Fig. 3.10 architecture, the legacy systems like IBIS/BRICS will be eliminated and replaced with more sophisticated and more powerful smart contracts inside Ethereum Blockchain to achieve the same or streamlined functions as IBIS. Data dictionary can become a set of smart contracts inside Blockchain. Form structure and data collection protocol can be realized via smart contracts and some kind of DAPP manager outside the Blockchain network. A new data analytic smart contract can be programmed to extract

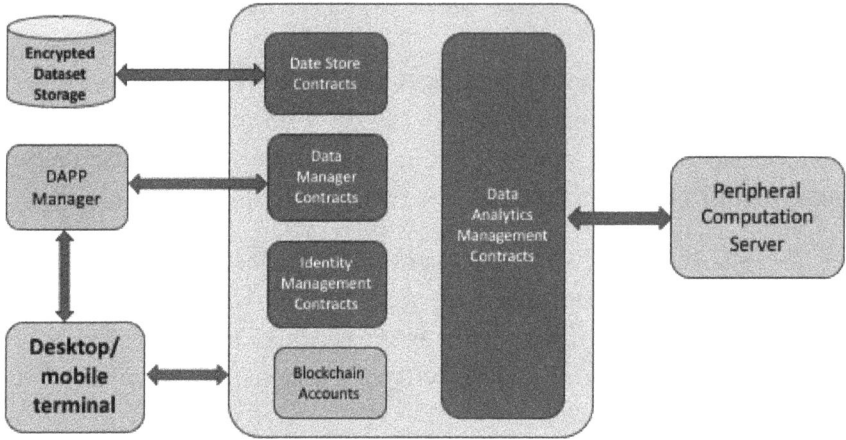

Fig. 3.10.  Blockchain-centric architecture for dataset management and big data analytics.

useful attributes from dataset for data analytics. A Computation Server outside Blockchain network can be used for conducting computation intensive work.

## 3.7.  Future Work

To eliminate a centralized system like IBIS/BRICS in the long run and replace with Decentralized Application (DAPP), dataflow built in IBIS/BRICS can be re-architected to run using Blockchain smart contracts. In conjunction of the development of Data Analytics contract and external computation server, Data Store contract needs to be extended to capture dataset features that are useful for Big Data analysis. Identity Management contract shall take a global management view of all Data Manager contracts in the Blockchain network.

## 3.8.  Summary

An informatics architecture is used to integrate the IBIS/BRICS with Ethereum Blockchain network to provide secure and reliable data for Big Data analysis. IBIS/BRICS was used to load Taiwan Stroke Registry data into an IBIS/BRICS data depository. During this research, data dictionary, user definable form structure, and data collection protocol built in the IBIS/BRICS data flow greatly benefit the import of Stroke data. Use of Mirth Connect and a new Blockchain Adapter to further streamline the dataset into Ethereum Blockchain network can create a data platform for more secure, reliable, transparent, and trustful dataset exchange among medical organizations. More importantly, with the increased trust on exchanged data over Blockchain network, Big Data analytics can be effectively performed in the future. In the long run, informatics system outside Blockchain network can be minimized (thus, further increase system security and reliability) or replaced with more sophisticated and powerful smart contracts inside Blockchain network.

## Acknowledgment

This research is partially supported by the Ministry of Science and Technology through Pervasive Artificial Intelligence Research (PAIR) Labs, Taiwan under the grants MOST 108-2634-F-468-001 and MOST 106-2632-E-468-002. The authors of this paper would also like to acknowledge Dr. Yang Fann for providing knowledge and development history of the IBIS. Dr. Fann is currently Director, Intramural IT and Bioinformatics Program at NIH. He co-led the development of IBIS/BRICS and informatics infrastructure named BRICS (Biomedical Research Informatics Computing System) to support and catalyze biomedical research and data sharing. In the last few years, he has shared the vision of IBIS/BRICS and helped promoting adoption and deployment of IBIS/BRICS in Taiwan.

## References

1. McAfee, A. and Brynjolfsson, E. (2012). Big data: the management revolution. *Harvard Business Review* Oct; 90(10), pp. 60–66, 68, 128.
2. Jacobs Adam. (2009). The pathologies of big data communications of the ACM. 52(8), pp. 36–44. doi: 10.1145/1536616.1536632.
3. Matthew J. McAuliffe, https://commonfund.nih.gov/sites/default/files/SPARCMcAuliffe-CIT-Sess2.pdf, NIH/CIT.

4. James Ray edited. A Next-Generation Smart Contract and Decentralized Application Platform, August 22, 2018. https://github.com/ethereum/wiki/wiki/White-Paper#ethereum.

5. Hsieh, F. I., Lien, L. M., Chen, S. T., Bai, C. H., Sun, M. C., Tseng, H. P., Chen, Y. W., Chen, C. H., Jeng, J. S., Tsai, S. Y., Lin, H. J., Liu, C. H., Lo, Y. K., Chen, H. J., Chiu, H. C., Lai, M. L., Lin, R. T., Sun, M. H., Yip, B. S., Chiou, H. Y. and Hsu, C. Y. (2010). Taiwan Stroke Registry Investigators. Get With the Guidelines-Stroke performance indicators: surveillance of stroke care in the Taiwan Stroke Registry. *Circulation* Sep 14; 122(11), pp. 1116–1123.

6. Furht, B. and Agarwal, A. (2013). *Handbook of Medical and Healthcare Technologies*. SpringerLink. Springer, New York. p. 291. ISBN 978-1-4614-8495-0. Retrieved December 5, 2018.

7. Trotter, F. and Uhlman, D. (2011). Meaningful Use and Beyond: A Guide for IT Staff in Health Care. Real Time Bks. O'Reilly Media. p. 223. ISBN 978-1-4493-0502-4. Retrieved December 5, 2018.

8. https://cnrm-dr.nih.gov/ and https://fitbir.nih.gov/.

9. Trygve Reenskaug and James Coplien. More deeply, the framework exists to separate the representation of information from user interaction. The DCI Architecture: A New Vision of Object-Oriented Programming — March 20, 2009.

10. Ben Alex and Luke Taylor. Spring Security 2.0 Reference Documentation, https://docs.spring.io/spring-security/site/docs/2.0.x/reference/html/springsecurity.html.

Chapter 4

# Analysis of Circulating Tumor DNA in Patients with Cancer: A Clinical Perspective

Chi-Chun Yeh and Peter Mu-Hsin Chang

*Jin An Clinic and Department of Oncology,*
*Taipei Veterans General Hospital,*
*Taipei, Taiwan*

## Abstract

Somatic genomic alterations are drivers of initiation, progression, and development of resistance to treatment of cancer. Analysis of circulating tumor DNA
(ctDNA) enables detection of somatic genomic alteration non-invasively with a
simple blood draw. ctDNA analysis may overcome the obstacles of spatial and
temporal genomic heterogeneity of cancer and may provide essential information to facilitate an informed clinical decision-making. Emerging evidence has
demonstrated clinical utility of ctDNA in directing treatment option, determining
prognosis of patients, detecting recurrence of cancer, and detecting early primary
cancer. Methods are developed to achieve detection of genomic alterations with
high sensitivity and specificity. A brief review of clinical applications of ctDNA
analysis is provided.

*Keywords*: cancer, cell-free DNA (cfDNA), circulating tumor DNA (ctDNA),
next-generation sequencing (NGS).

## 4.1. Introduction

Cancers evolve [1]. Somatic genomic alterations those confer fitness advantages to cancer cells harbor them to drive the growth of

cancer [2]. Large cancer sequencing projects have depicted landscapes of somatic genomic alteration for major cancer types [3–11]. The circulating tumor DNA (ctDNA) analysis technology enables observation of *in vivo* cancer evolution "in action" with serial blood sampling [12]. In this chapter, we will start with a clinical scenario demonstrating the current clinical application of ctDNA analysis in directing treatment option, followed by a brief technological highlight of ctDNA analysis and emerging evidence demonstrating the clinical utility of ctDNA analysis.

## 4.2.  Clinical Scenario

A 41-year-old male had been diagnosed with adenocarcinoma of lung; stage IV with lung-to-lung metastasis and hepatic metastasis since 20 months ago. The pulmonary tumor tissue obtained by CT-guided transthoracic biopsy 20 months ago was positive for the $EGFR^{L858R}$ mutation. Treatment with erlotinib, an oral EGFR tyrosine kinase inhibitor, was initiated since 19 months ago. The follow-up CT scans during treatment with erlotinib demonstrated that both pulmonary and hepatic tumors responded partially to erlotinib treatment. However, the follow-up CT scan performed 7 months ago demonstrate that pulmonary and hepatic tumors progressed. The patient developed dyspnea and hemoptysis. A ctDNA analysis assay was performed with 10 ml of blood sample 6 months ago. The ctDNA analysis assay demonstrated $EGFR^{L858R}$ (mutant allele fraction: 15%) and $EGFR^{T790M}$ (mutant allele fraction: 8.5%). Under the impression of acquired resistance to erlotinib mediated by $EGFR^{T790M}$ mutation, treatment with osimertinib, a third-generation EGFR, was initiated since 6 months ago. The pulmonary and hepatic tumors responded to osimertinib treatment partially as demonstrated by follow-up CT scan performed 3 months ago. The patient remained on osimertinib treatment until now.

## 4.3.  Current Knowledge about Molecular Mechanisms of Acquired Resistance to Treatment

In this common clinical scenario, a cancer was genomically profiled and genomic alteration-targeted treatment was given accordingly and cancer responded to the genomic alteration-targeted treatment. However, the initially responsive tumor progressed again, a phenomenon called "acquired resistance" to cancer treatment [13–15]. A thorough understanding of the mechanisms of acquired resistance and knowledge about proper detection, evaluation, and management of acquired resistance is essential to benefit patients with cancer.

Cancer is a genomic disease. The initiation, progression, and development of resistance to treatment of cancer are driven by somatic alteration of the genome that confers a fitness advantage to the cancer cell [2]. The somatic genomic alteration (SGA) that drives cancer is called "driver SGA" [2]. A driver SGA occurred on the gene that plays a central role in the regulatory circuit that controls proliferation, survival, differentiation, and metabolisms of the cell [16]. These genes are called "cancer gene". Several driver SGAs act in concert to convert normal cells into malignant ones [17, 18]. Cancer cells tend to de-differentiate and abandon their physiological function, are endowed with increased proliferation capacity, enhanced survival, and re-wire their metabolic state to sustain an unrestrained growth of the tumor [17, 18]. Modern cancer genome sequencing project has cataloged human cancer genes [2]. Acknowledging the cancer gene and driver SGA as an engine for cancer growth, the therapeutic strategy that inhibits the molecular function of cancer gene and driver SGA has started to gain initial success. For example, the EGFR TKI erlotinib or gefitinib yielded a response rate of about 60–70% in patients with non-small cell lung cancer with sensitizing EGFR mutation [19–21]. However, most, if not all, patients developed acquired resistance to erlotinib and gefitinib after a treatment period of about 9–10 months [19–21]. The genomic mechanism that underlies the development of acquired resistance to EGFR TKI was studied extensively. The most common genomic mechanism underlies acquired resistance to EGFR TKI is the occurrence of $EGFR^{T790M}$ mutation [22]. Other mechanisms of acquired resistance to EGFR TKI include high-copy number MET amplification, EGFR amplification, and transformation to small cell carcinoma [23]. The therapeutic strategy to overcome $EGFR^{T790M}$-mediated has been developed. Osimertinib, a third-generation EGFR TKI that selectively inhibits sensitizing EGFR mutation and $EGFR^{T790M}$, has demonstrated a response rate of about 70% in patients with non-small cell lung cancer with the EGFR$^{T790M}$ mutation [24]. The clinical development of EGFR-targeted treatment in non-small cell lung cancer has demonstrated that a better understanding of genomic underpinning of cancer can be translated to a better cancer care.

An evolutionary viewpoint can provide deep insight into mechanisms of acquired resistance to cancer treatment. Evolutionary forces, such as selection, genetic drift, mutation, and migration, shape initiation, progression, and development of drug resistance of cancers [25, 26]. Cancer evolution is a dynamical process that involves complex interactions between cancers and their microenvironment [27]. SGAs accumulate through every cycle of cell replication. Moreover, cancers those develop mutator phenotype accumulate SGAs at a higher rate [28]. A study that uses multi-region sequencing to estimate the burden of SGAs within a single tumor from a patient with hepatocellular carcinoma demonstrates there is more than 100 millions coding-region alteration within a single tumor [29]. Selective pressures, such as anti-cancer immunity and anti-cancer treatment act upon SGAs [27]. SGAs those confer a selective advantage to cancer cells will drive the growth of cancer [27]. Thus, an assay that detects SGAs non-invasively with blood sample may help in depicting evolutionary trajectories of cancers.

## 4.4. Analysis of ctDNA to Detect Somatic Alterations of Genome That Underlie the Development of Resistance to Treatment

Cell-free nucleic acid, including cell-free DNA (cfDNA), cell-free mRNA, and cell-free miRNA, are small nucleic acid fragments released into the bloodstream from necrotic, apoptotic, or exocytotic cells that originate from all kinds of cells within the body [30]. Cell-free tumor-derived DNA (ctDNA) is a fraction of cfDNA from a primary or metastatic tumor. The half-life of cell-free nucleic acid in the bloodstream ranges from 15 mins to several hours [31]. A well-established clinical application of cfDNA analysis is to detect fetal aneuploidy by analyzing fetal DNA that circulates within the maternal circulation [32]. In line with using fetal DNA in maternal blood to interrogate genomic makeup of fetus, ctDNA analysis can be used to analyze genomic alterations of cancer [33]. Major advantages of ctDNA analysis over tissue-based examination of tumor DNA is that ctDNA analysis is able to overcome obstacles of spatial and temporal intra-tumoral heterogeneity of genomic alterations [34]. In respect of spatial heterogeneity of tumor, the genomic alterations discovered by ctDNA analysis are a global summary of alterations derived from heterogeneous cancer cells both from primary and metastatic sites within a subject [35–38]. On the contrary, tissue-based examinations derive tumor DNA from a specific area of the tumor, may overlook genomic alterations occurring outside the biopsied area [39]. As for temporal heterogeneity of genomic alteration, cancers evolve through time, so genomic alterations change dynamically over time [40]. So genomic alterations reported by analyzing archived tumor tissue at diagnosis may be less informative when the molecular mechanism of acquired resistance to treatment is needed for proper decision-making. To uncover "current" genomic alterations those are most relevant to clinical decision-making [41], one needs an assay that can report genomic alterations of a tumor at any time point as dictated by clinical conditions. A tumor tissue can be obtained only by invasive procedures such as surgical excision or biopsy. However, surgical excision or biopsy is painful, risky, costly, and are not always permitted by the clinical situation [42]. So tumor tissues are difficult, if not impossible, to obtain at several time points during clinical course. The ctDNA analysis uses blood samples, which can be obtained with a simple blood draw without concern of risk of the procedure. To overcome temporal heterogeneity of tumor, the ctDNA analysis can be performed at any time points to report molecular mechanisms those are most important for clinical care [43–46].

## 4.5. Method to Increase Analytical Specificity and Sensitivity of ctDNA Analysis

The mutant allele fraction (MAF) is defined as the fraction of mutant DNA allele in the total DNA allele found in cfDNA. The MAF is used to indicate relative concentration of a genomic alteration within the blood. It has been reported that MAF varies markedly among patients with cancer. In a series of more than 15,000 cases of cancers of multiple histologic types, 25th percentile, median, 75th percentile of MAF of reported somatic genomic alterations are 0.18%, 0.39%, and 1.94% respectively [47]. The fact that somatic genomic alterations from tumor circulating within blood at relatively low concentration poses an analytical challenge of the ctDNA analysis. The ctDNA analysis is based on the PCR and NGS technologies. However, the intrinsic error arising from PCR and NGS workflow prevents confident detection of somatic genomic alteration of low MAF [48]. Several teams have developed methods to modify NGS workflow to achieve high sensitivity and specificity of detection of somatic genomic alteration even at low MAF [49–55]. There are several variations of the method to improve analytical specificity and sensitivity of ctDNA analysis. However, the essential concepts of these methods are similar. Basically, before sequencing, individual DNA molecules are tagged with unique identifiers to enable post-sequencing tracking of individual DNA molecule and reconstruction of consensus sequence of each original DNA molecule. Thus, sequencing errors can be distinguished from authentic mutations. By applying above methods, per base analytical specificity of detection of genomic alterations of >99.9999% and analytical sensitivity down to 0.1% MAF can be achieved [55].

### 4.6.  Clinical Significance of Genomic Alterations Detected by ctDNA Analysis

In addition to analytical validation, the clinical significance of genomic alterations detected by ctDNA analysis should be determined to implement this technology in clinical settings. To understand the clinical significance of genomic alterations detected by ctDNA analysis, testing results of ctDNA analysis should correlate with data of patient outcomes obtained from well-designed clinical studies. Several potential clinical applications of ctDNA analysis have been explored, including detecting molecular mechanisms of acquired resistance to direct treatment option [56], determining prognosis [57], detection of recurrence of cancer [58–61], and early detection of primary cancer [62–64]. Although immature, evidence demonstrating clinical significance of ctDNA analysis is starting to emerge.

In a cohort of patients with $EGFR^{T790M}$ (+) adenocarcinoma of lung treated with osimertinib in the second-line setting, patients with positive testing results of EGFR$^{T790M}$ by ctDNA analysis by PCR-based method responded comparably to patients with positive testing results by analyzing tumor tissue, demonstrating the potential of ctDNA analysis in directing treatment option [56]. Moreover, patients with positive testing results of EGFR$^{T790M}$ with tissue-based methods have a favorable response to osimertinib treatment and longer progression-free survival, demonstrating the value of ctDNA analysis in determining the prognosis of patients [56].

In a cohort of early lung cancer treated with surgery and post-surgery adjuvant chemotherapy, the ctDNA analysis was performed before surgery and regularly at a 3-month interval after surgery for the first 2 years following enrollment and every 6 months thereafter. Serial post-surgery ctDNA analysis reported a median lead time between detectable ctDNA and image study-confirmed recurrence of 70 days, indicating ctDNA analysis may enable early detection of recurrence of cancer [61].

An enhanced ability to detect cancer earlier, when most of the cancers can be cured with surgical treatment, will greatly decrease mortality from cancer [65, 66]. An assay for the early detection of cancer needs to have high sensitivity and specificity. A false negative testing result may lead to under-diagnosis and under-treatment of cancer. A false positive testing result may lead to over-diagnosis and over-treatment. ctDNA analysis, which measures signals directly from cancer cells with high analytical sensitivity and specificity, holds great promises in the domain of early detection of cancer. A huge clinical project, the circulating cancer genome atlas (CCGA) study, aiming to deep sequence cfDNA of 15,000 subjects prospectively (4,500 non-cancer subjects and 10,500 cancer subjects) and follow-up for up to 5 years, is undergoing to establish genomic landscapes of cfDNA of non-cancer and cancer subjects [67]. To discover trivial differences in genomic landscapes between non-cancer and cancer subjects, a large subject population size and a long follow-up period are needed. Initial data, when 1,627 subjects, 749 non-cancer control and 878 cancer subjects were included for analysis, demonstrated that sensitivity of detection of stage I–III cancers by whole–genome bisulfite sequencing of cfDNA is 66% for colorectal cancer, 77% for lymphoma, 73% for multiple myeloma, 90% for ovarian cancer, and 80% for pancreatic cancer [63]. Cancers of the low sensitivity of detection (less than 10%) include low Gleason score prostate cancer, thyroid cancer, uterine cancer, melanoma, and renal cancer [63]. The clinical significance of detected genomic alterations in cfDNA in non-cancer and cancer subjects may be better understood upon completion of the CCGA study.

## 4.7. Conclusion

The ctDNA analysis could provide deep insights into the initiation, progression, and development of resistance to treatment of cancers. Genomic information of tumor, which is important to clinical decision-making, can be obtained with a non-invasive blood draw. Through well-designed clinical studies, these biological insights can be translated into clinical intelligence. Emerging evidence has demonstrated the clinical utility of ctDNA analysis in directing treatment option, determining the prognosis of patients, detecting recurrence, and detecting primary cancer. Although immature, these clinical applications of ctDNA analysis have demonstrated the potential to drastically transform the current practice of oncology. Furthermore, rapid technological advancement in the generation, analysis, and storage of sequencing data will accelerate widespread adoption of sequencing-based assays in the oncology clinics.

## References

1. Greaves, M. and Maley, C. C. (2012). Clonal evolution in cancer. *Nature* 481, pp. 306–313.
2. Bailey, M. H., Tokheim, C., Porta-Pardo, E., Sengupta, S., Bertrand, D., Weerasinghe, A., Colaprico, A., Wendl, M. C., Kim, J., Reardon, B. *et al.* (2018). Comprehensive characterization of cancer driver genes and mutations. *Cell*, 17310.1016/j.cell.2018.02.060.
3. Cancer Genome Atlas, N. (2012). Comprehensive molecular characterization of human colon and rectal cancer. *Nature* 487, pp. 330–337.
4. Comprehensive molecular portraits of human breast tumours. (2012). *Nature* 490(7418), pp. 61–70. Epub 2012/09/25. pmid:23000897.
5. Comprehensive genomic characterization of squamous cell lung cancers. (2012). *Nature* 489(7417), pp. 519–525. Epub 2012/09/11. pmid:22960745.

6. Cancer Genome Atlas Research, N. (2013). Genomic and epigenomic landscapes of adult de novo acute myeloid leukemia. *The New England Journal of Medicine* 368, pp. 2059–2074.
7. Brat, D. J., Verhaak, R. G., Aldape, K. D., Yung, W. K., Salama, S. R., Cooper, L. A., Rheinbay, E., Miller, C. R., Vitucci, M., Morozova, O. *et al.* (2015). Cancer genome atlas research network. Comprehensive, integrative genomic analysis of diffuse lower-grade gliomas. *The New England Journal of Medicine* 372, pp. 2481–2498.
8. Cancer Genome Atlas Research Network (2011). Integrated genomic analyses of ovarian carcinoma. *Nature* 474, pp. 609–615.
9. Cancer Genome Atlas Research Network (2014). Comprehensive molecular characterization of urothelial bladder carcinoma. *Nature* 507, pp. 315–322.
10. Cancer Genome Atlas Research Network (2013). Comprehensive molecular characterization of clear cell renal cell carcinoma. *Nature* 499, pp. 43–49.
11. Brennan, C. W. *et al.* (2013). The somatic genomic landscape of glioblastoma. *Cell* 155, pp. 462–477.
12. Webb, S. (2016). The cancer bloodhounds. *Nature Biotechnology* 34, pp. 1090–1094. https://doi.org/10.1038/nbt.3717.
13. Ahn, I. E., Underbayev, C., Albitar, A. *et al.* (2017). Clonal evolution leading to ibrutinib resistance in chronic lymphocytic leukemia. *Blood* 129, pp. 1469–1479.
14. Landau, D. A., Carter, S. L., Stojanov, P. *et al.* (2013). Evolution and impact of sub-clonal mutations in chronic lymphocytic leukemia. *Cell* 152, pp. 714–726.
15. Jackman, D., Pao, W., Riely, G. J. *et al.* (2010). Clinical definition of acquired resistance to epidermal growth factor receptor tyrosine kinase inhibitors in non-small-cell lung cancer. *Journal of Clinical Oncology* 28(2), pp. 357–360.
16. Futreal, P. A. *et al.* (2004). A census of human cancer genes. *Nature Reviews Cancer* 4, pp. 177–183.
17. Hanahan, D. and Weinberg, R. A. (2000). The hallmarks of cancer. *Cell* 100, pp. 57–70.
18. Hanahan, D. and Weinberg, R. A. (2011). Hallmarks of cancer: The next generation. *Cell* 144, pp. 646–674.
19. Rosell, R., Carcereny, E., Gervais, R., Vergnenegre, A., Massuti, B., Felip, E., Palmero, R., Garcia-Gomez, R., Pallares, C., Sanchez, J. M., Porta, R., Cobo, M., Garrido, P. *et al.* (2012). Erlotinib versus standard chemotherapy as first-line treatment for European patients with advanced EGFR mutation-positive non-small-cell lung cancer (EURTAC): a multicentre, open-label, randomised phase 3 trial. *Lancet Oncology* 13, pp. 239–246.
20. Zhou, C., Wu, Y. L., Chen, G., Feng, J., Liu, X. Q., Wang, C., Zhang, S., Wang, J., Zhou, S., Ren, S., Lu, S., Zhang, L., Hu, C. *et al.* (2011). Erlotinib versus chemotherapy as first-line treatment for patients with advanced EGFR mutation-positive non-small-cell lung cancer (OPTIMAL, CTONG-0802): a multicentre, open-label, randomised, phase 3 study. *Lancet Oncology* 12, pp. 735–742.
21. Mok, T. S., Wu, Y.-L., Thongprasert, S. *et al.* (2009). Gefitinib or carboplatin-paclitaxel in pulmonary adenocarcinoma. *The New England Journal of Medicine* 361, pp. 947–957.
22. Yun, C. H., Mengwasser, K. E., Toms, A. V. *et al.* (2008). The T790M mutation in EGFR kinase causes drug resistance by increasing the affinity for ATP. *Proceedings of the National Academy of Sciences of the United States of America* 105, pp. 2070–2075.

23. Yu, H. A., Arcila, M. E., Rekhtman, N. *et al.* (2013). Analysis of tumor specimens at the time of acquired resistance to EGFR-TKI therapy in 155 patients with EGFR-mutant lung cancers. *Clinical Cancer Research* 19, pp. 2240–2247.

24. Mok, T. S., Wu, Y.-L., Ahn, M.-J. *et al.* (2017). Osimertinib or platinum–pemetrexed in EGFR T790M–positive lung cancer. *The New England Journal of Medicine* 376, pp. 629–640.

25. Merlo, L. M., Pepper, J. W., Reid, B. J. and Maley, C. C. (2006). Cancer as an evolutionary and ecological process. *Nature Reviews Cancer* 6, pp. 924–935.

26. Stratton, M. R., Campbell, P. J. and Futreal, P. A. (2009). The cancer genome. *Nature* 458, pp. 719–724.

27. Maley, C. C. *et al.* (2017). Classifying the evolutionary and ecological features of neoplasms. *Nature Reviews Cancer* 17, pp. 605–619.

28. Loeb, L. A., Bielas, J. H., Beckman, R. A. and Bodmer, I. W. (2008). Cancers exhibit a mutator phenotype: clinical implications. *Cancer Research* 68, pp. 3551–3557.

29. Ling, S., Hu, Z., Yang, Z., Yang, F., Li, Y., Lin, P., Chen, K., Dong, L., Cao, L., Tao, Y. *et al.* (2015). Extremely high genetic diversity in a single tumor points to prevalence of non-Darwinian cell evolution. *Proceedings of the National Academy of Sciences of the United States of America* 112, pp. E6496–E6505.

30. Schwarzenbach, H., Hoon, D. S. and Pantel, K. (2011). Cell-free nucleic acids as biomarkers in cancer patients. *Nature Reviews Cancer* 11, pp. 426–437.

31. Fleischhacker, M. and Schmidt, B. (2007). Circulating nucleic acids (CNAs) and cancer-a survey. *Biochimica et Biophysica Acta* 1775, pp. 181–232.

32. Lo, Y. M., Corbetta, N., Chamberlain, P. F. *et al.* (1997). Presence of fetal DNA in maternal plasma and serum. *Lancet* 350, pp. 485–487.

33. Bettegowda, C., Sausen, M., Leary, R. J., Kinde, I., Wang, Y., Agrawal, N., Bartlett, B. R., Wang, H., Luber, B., Alani, R. M. *et al.* (2014). Detection of circulating tumor DNA in early- and late-stage human malignancies. *Science Translational Medicine* 6, p. 224ra224.

34. Diaz, L. A. and Bardelli, A. (2014). Liquid biopsies: genotyping circulating tumor DNA. *Journal of Clinical Oncology* 32, pp. 579–586.

35. Murtaza, M. *et al.* (2015). Multifocal clonal evolution characterized using circulating tumour DNA in a case of metastatic breast cancer. *Nature Communications* 6, p. 8760.

36. Kuo, Y.-B., Chen, J. S., Fand, C. W., Li, Y. S. and Chan, E. C. (2014). Comparison of KRAS mutation analysis of primary tumors and matched circulating cell-free DNA in plasmas of patients with colorectal cancer. *Clinica Chimica Acta* 433, pp. 284–289.

37. De Mattos-Arruda, L. *et al.* (2014). Capturing intra-tumor genetic heterogeneity by de novo mutation profiling of circulating cell-free tumor DNA: A proof-of-principle. *Annals of Oncology* 25, pp. 1729–1735.

38. Jamal-Hanjani, M. *et al.* (2016). Detection of ubiquitous and heterogeneous mutations in cell-free DNA from patients with early-stage non-small-cell lung cancer. *Annals of Oncology* 27, pp. 862–867.

39. Gerlinger, M. *et al.* (2012). Intratumor heterogeneity and branched evolution revealed by multiregion sequencing. *The New England Journal of Medicine* 366, pp. 883–892.

40. Swanton, C. (2012). Intratumor heterogeneity: Evolution through space and time. *Cancer Research* 72, pp. 4875–4882.

41. Diaz, L. A. *et al.* (2012). The molecular evolution of acquired resistance to targeted EGFR blockade in colorectal cancers. *Nature* 486, pp. 537–540.
42. Wu, C. C., Maher, M. M. and Shepard, J. A. (2011). Complications of CT-guided percutaneous needle biopsy of the chest: Prevention and management. *American Journal of Roentgenology* 196, pp. W678–W682.
43. Frenel, J.-S. *et al.* (2015). Serial next generation sequencing of circulating cell free DNA evaluating tumour clone response to molecularly targeted drug administration. *Clinical Cancer Research* 21, pp. 4586–4596.
44. Siravegna, G. *et al.* (2015). Clonal evolution and resistance to EGFR blockade in the blood of colorectal cancer patients. *Nature Medicine* 21, pp. 795–801.
45. Misale, S. *et al.* (2012). Emergence of KRAS mutations and acquired resistance to anti-EGFR therapy in colorectal cancer. *Nature* 486, pp. 532–536.
46. Thress, K. S. *et al.* (2015). Acquired EGFR C797S mutation mediates resistance to AZD9291 in non-small cell lung cancer harboring EGFR T790M. *Nature Medicine* 21, pp. 560–562.
47. Zill, O. A. *et al.* (2016). Somatic genomic landscape of over 15,000 patients with advanced-stage cancer from clinical next-generation sequencing analysis of circulating tumor DNA [abstract]. *Journal of Clinical Oncology* 34(Suppl.), p. LBA11501.
48. Fox, E. J., Reid-Bayliss, K. S., Emond, M. J. and Loeb, L. A. (2014). Accuracy of next generation sequencing platforms. *Journal of Next Generation Sequencing and Applications* 1, p. 1000106.
49. Kinde, I., Wu, J., Papadopoulos, N., Kinzler, K. W. and Vogelstein, B. (2011). Detection and quantification of rare mutations with massively parallel sequencing. *Proceedings of the National Academy of Sciences of the United States of America* 108, pp. 9530–9535.
50. Schmitt, M. W. *et al.* (2012). Detection of ultra-rare mutations by next-generation sequencing. *Proceedings of the National Academy of Sciences of the United States of America* 109, pp. 14508–14513.
51. Kennedy, S. R. *et al.* (2014). Detecting ultralow-frequency mutations by Duplex Sequencing. *Nature Protocols* 9, pp. 2586–2606.
52. Gregory, M. T. *et al.* (2016). Targeted single molecule mutation detection with massively parallel sequencing. *Nucleic Acids Research* 44, p. e22.
53. Kukita, Y. *et al.* (2015). High-fidelity target sequencing of individual molecules identified using barcode sequences: de novo detection and absolute quantitation of mutations in plasma cell-free DNA from cancer patients. *DNA Research* 22, pp. 269–277.
54. Schmitt, M. W. *et al.* (2015). Sequencing small genomic targets with high efficiency and extreme accuracy. *Nature Methods* 12, pp. 423–425.
55. Lanman, R. B. *et al.* (2015). Analytical and clinical validation of a digital sequencing panel for quantitative, highly accurate evaluation of cell-free circulating tumor DNA. *PLoS ONE* 10, e0140712.
56. Oxnard, G. R. *et al.* (2016). Association between plasma genotyping and outcomes of treatment with osimertinib (AZD9291) in advanced non-small-cell lung cancer. *Journal of Clinical Oncology* 34, pp. 3375–3382.
57. Stover, D. G., Parsons, H. A., Ha, G. *et al.* (2018). Association of cell-free DNA tumor fraction and somatic copy number alterations with survival in metastatic triple-negative breast cancer. *Journal of Clinical Oncology* 36, pp. 543–553.

58. Garcia-Murillas, I. *et al.* (2015). Mutation tracking in circulating tumor DNA predicts relapse in early breast cancer. *Science Translational Medicine* 7, p. 302ra133.
59. Reinert, T. *et al.* (2015). Analysis of circulating tumour DNA to monitor disease burden following colorectal cancer surgery. *Gut* 65, pp. 625–634.
60. Olsson, E. *et al.* (2015). Serial monitoring of circulating tumor DNA in patients with primary breast cancer for detection of occult metastatic disease. *EMBO Molecular Medicine* 7, pp. 1034–1047.
61. Abbosh, C. *et al.* (2017). Phylogenetic ctDNA analysis depicts early-stage lung cancer evolution. *Nature* 545, pp. 446–451.
62. Oxnard, G. R., Maddala, T., Hubbell, E. *et al.* (2018). Genome-wide sequencing for early stage lung cancer detection from plasma cell-free DNA (cfDNA): The Circulating Cancer Genome Atlas (CCGA) study. *Journal of Clinical Oncology* 36(suppl; abstr LBA8501).
63. Klein, E. A., Hubbell, E., Maddala, T. *et al.* (2018). Development of a comprehensive cell-free DNA (cfDNA) assay for early detection of multiple tumor types: The Circulating Cell-free Genome Atlas (CCGA) study. *Journal of Clinical Oncology* 36(suppl; abstr 12021).
64. Liu, M. C., Maddala, T., Aravanis, A. *et al.* (2018). Breast cancer cell-free DNA (cfDNA) profiles reflect underlying tumor biology: The Circulating Cell-Free Genome Atlas (CCGA) study. *Journal of Clinical Oncology* 36(suppl; abstr 536).
65. Cho, H., Mariotto, A. B., Schwartz, L. M., Luo, J. and Woloshin, S. J. (2014). *National Cancer Institute Monograph* 2014, pp. 187–197.
66. Aravanis, A. M., Lee, M. and Klausner, R. D. (2017). Next-generation sequencing of circulating tumor DNA for early cancer detection. *Cell* 168, pp. 571–574.
67. ClinicalTrials.gov Identifier: NCT02889978.

Chapter 5

# Big Data Computation of Drug Design: From the Natural Products to the Transcriptomic-Based Molecular Development

David Agustriawan, Arli Aditya Parikesit,
and Rizky Nurdiansyah

*Department of Bioinformatics,*
*Indonesia International Institute for Life Sciences,*
*Indonesia*

## 5.1. Introduction

The field of biology is entering the era of big data after the invention of high-throughput technologies, such as microarray and next-generation sequencing. Omics fields also emerge, as the technology permits, and it steadily produces data, with almost 1 zettabyte of data per year [1a]. Those enormous pile of data changes the way scientist of analysis to extract information as much as possible [1a; Dai *et al.*, 2012]. However, it also opens new doors and perspective to solve biological problems, such as drug discovery. This particular chapter will discuss some recent advances of how big data can help in drug discovery for some diseases, managing the natural resource knowledge, and supports the decision-making in the process of drug design.

HIV-1 (Human Immunodeficiency Virus type-1) and Alzheimer disease (AD) treatments are still needed to be developed since the current drug do not have a significant effect to cure that disease. HIV-1 is a member of the retrovirus family that causes AIDS in human [1a]. The epidemic of AIDS is considered as one of the most destructive diseases and it will make the immune system weakened and caused to lethal opportunistic infection [2, 3]. There were more than 33 million people infected by HIV until 2010 and this amount of number will increase approximately by 2–3 million per year [4]. On the other hand, AD is the most common late-life neurodegenerative disorder that affects approximately 36 million people worldwide as of 2009 [5]. There is an enormous amount of evidence that AD is associated with oligomerization of beta-amyloid (A$\beta$) peptides.

Currently, both HIV-1 and AD treatments are not efficient to treat those diseases. For example, there are five drugs; Donepezil, Galantamine, Rivastigmine, Tacrine, and Memantine; that are prescribed for symptomatic treatments of AD. The first four medicines are acetylcholinesterase inhibitors, whereas memantine modulates N-Methyl-D-aspartic acid. Lately, doctors rarely prescribe Tacrine because it is associated with more serious side effects than the other drugs. Those drugs does not slow down or reverse the progression of AD disease, and that problem also happened in the AIDS treatment. Therefore, the search for new leads is of great interest. Treatment development from natural products can be one of the alternatives as a new treatment for AD and HIV-1. Natural products are chemical compounds resulting from the evolution process in plants, marine organisms, and fungi [6]. Many such chemicals have been produced and optimized under prey/predator selective coevolution forces. Such natural compounds have been utilized by humans since ancient times for treatments and cures of many diseases. From all 40,000 species of plants on earth, 30,000 species of which live in Indonesia archipelago, and 9,600 of that 30,000 species have pharmacology activities [7].

In a wet-lab, particularly *in vivo* or *in vitro* analysis, screening of pharmacological activity of active ingredients in medicinal plants requires an expensive process, energy, qualified human resources, and not a short time [8].

*In silico* study is an attractive chance by the use of computers as tools in drug development. An exponential increase in computing capabilities will provide opportunities to develop simulations and calculations in drug design. Molecular docking and molecular dynamics study can be performed to create a simulation to predict an intermolecular complex between the drug molecule with its target protein. In addition to health risks for clinical trial participants, late failures are extremely costly as large amounts of time and capital have already been invested in developing the drug. Identifying the future failure early, even before they enter clinical phases, can save significant expenditure later on. The decisions during lead optimization are crucial because they determine which compounds will enter costly preclinical and clinical development (Table 5.1) [9].

One of the multidimensional assays that have gained considerable attention in the past decade is gene expression profiling. This technique simultaneously measures many biological effects of a compound on the transcriptional level and thereby gives a comprehensive snapshot of the biological state of the living system [10–12]. Transcriptomic changes following compound administration can now also be measured in high throughput, enabling screening of many compounds in multiple cell lines at low cost. The use of transcriptomic data for characterizing biological effects of small molecules has become increasingly popular since the advent of the connectivity map [13]. Several applications ranging from pathway elucidation [14], toxicity models [15, 16] and toxicogenomic classifications [17] to tool discovery and drug repurposing [18–21] have been developed based on drug-induced gene expression profiling [22]. This study also discusses the transcriptional effects of 757 compounds on eight cell lines using a total of ~1600 microarrays developed by Affymetrix [23]. It shows gene expression profiling to be a highly valuable tool for lead optimization in pharmaceutical discovery project.

## 5.2. The Conceptual Framework of this Study

This study will discuss *in silico* study in drug discovery, particularly for natural products compounds. This study will review molecular docking,

Table 5.1.  Typical decision points in drug discovery projects and the type of decision to be taken in each step [23].

| Decision point | Important criteria for decision | Decision support available |
|---|---|---|
| Choice of disease | Patient need; commercial aspects | Statistics on disease distributions; input from practitioners |
| Target selection | Validated target (i.e. involved in disease modulation and druggable) | Biological studies (e.g. knockdown experiments, genetic linkages); chemical biology/probes |
| Screening library assembly | Chemistry with no obvious liabilities, ease of synthesis of analogs, well assumed or proven PK/PD properties | Cheminformatics analysis of chemical space; historical hit distributions in chemical space |
| Assay development | Predictivity; reproducibility; throughput; price | Experience of biologists |
| Screening/hit list triaging | Data quality; increasing certainty about true and false positives and negatives | Experience of screeners/follow-up scientists |
| Lead optimization | On-target and off-target activities; favorable drug metabolism and pharmacological properties | Biochemical and more-complex assay systems; gene expression arrays |
| Preclinical studies | Efficacy and side-effect profile | Animal experiments |
| Clinical studies | Efficacy and side-effect profile | Testing of drug candidate in large (or stratified) cohorts |
| Approval | Efficacy and side-effect | Results from preclinical and clinical studies |
| Marketing | A market structure in the disease area; the comparative advantage of the drug with competitors in the market | Commercial information systems |

molecular dynamic studies in a natural product, and transcriptomic role as a tool in early-stage drug discovery decision making. As mentioned in Figure 5.1, the study is divided into three parts as follows:

  I. Molecular docking utilizes medicinal plants in Indonesia as AIDS therapy:

- Medicinal plants database and 3D structure of the chemical compounds from medicinal plants in Indonesia [24].
- Virtual screening of Indonesia herbal database as HIV-1 reverse transcriptase inhibitor [1a].

 II. Molecular docking and dynamics analysis of Vietnam herbal to treat Alzheimer:

- Top-leads from natural products for the treatment of Alzheimer's disease: docking and molecular dynamics study [25].

III. Transcriptomic role as a tool in early-stage drug discovery decision making:

- Using transcriptomics to guide lead optimization in drug discovery projects: Lesson learned from QSTAR project [23].

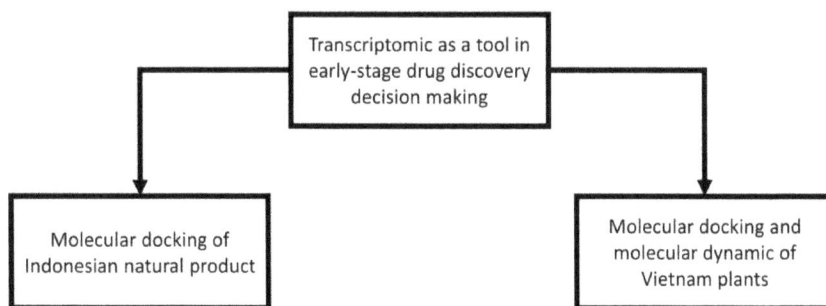

Fig. 5.1. Conceptual framework of the study.

## 5.3. Study Design

### 5.3.1. *Study I*: *Molecular Docking Utilizes Medicinal Plants in Indonesia as AIDS Therapy*

First, this study builds a database for medicinal plants and its natural compounds in Indonesia [24]. The website was set up and can be accessed at this following link http://herbaldb.farmasi.ui.ac.id. This study provided a 3D structure of the active compounds from medicinal plants database that can be used as input of molecular docking study. Then, *in silico* virtual screening approach was used to find lead molecules from the natural compound library or database as HIV-1 reverse transcriptase inhibitors [1a].

#### 5.3.1.1. *Material and method*

This research is done using a combination of literature study, generating the 3D structure of the medicinal plant, and from the generated structure, molecular docking was performed to find its target protein.

##### 5.3.1.1.1. Generating the 3D structure and preparation of medicinal plants database and websites

First, this study collected the information of medical plants in Indonesia through literature such as scientific journals, books, and websites. Then, the plants were sampled and kept. The data were taken from classic and official books [26–33]. After that, data on chemical compounds found in the medicinal plants along with its 2D structure were collected. 2D structure of the search query is performed on the PubChem chemical compound database [34] or KNApSAcK [35]. The output format from the KNApSAcK metabolite database is a gif file format, and then converted by drawing 2D structures of chemical compounds using Symyx Draw program. It yields .mol format [36]. Then, in order to generate a form of the 3D molecule, .mol or .sdf files, which have been downloaded from PubChem and converted with PyMOL [37], were used as input in the VEGAZZ program. After that convert the 2D structure to 3D structure with the said program. After obtaining the 3D structure, the mol format data then converted into .mol2 format using the OpenBabel program [38]. Lastly, this study established a web database prototype of medicinal plants and 3D structures of chemical compounds by utilizing MySQL software and PHP.

**Preparing the HIV-1 RT protein structure.** The HIV-1 reverse transcriptase structure consists of p66 and p51 subunit was obtained from protein databank websites [39]. The inclusion criteria were wild-type HIV-RT, binding with ligand or inhibitor, complete chain information and has a resolution less than 2.5 Å. Then the protein structure of HIV-1 RT bound with NVP (nevirapine) (Protein Data Bank (PDB) ID: 3LPI) was obtained and optimized using VEGAZZ [40, 41]. The optimization includes the addition of hydrogen with "protein" and "each residue ends" option, separation of solvent molecules and cofactor or ligand continued by adding partial charges (Gasteiger) and applying AutoDock forcefield with AutoDockTools in MGLTools 1.5.4.

**Preparing ligand file format.** The 3D structure of medicinal plants in the previous step was then optimized by adding hydrogen. Furthermore, the minimization was done using steepest descent and conjugate gradient 1,000 steps for each method. The last step was adding Gasteiger partial charges and applying AutoDock force field with AutoDockTools.

**Validating of molecular docking protocol.** In the case of 3LP1, the ligand binding site coordinates were defined by AutoDockTools. This binding site will be used as the center target of molecular docking for the virtual screening. The coordinates of binding site were $x = 10.350$, $y = 14.076$; and $z = 18.252$ in NNRTI pocket. Preliminary docking was done to validate the protocol using 14 control compounds given in Table 5.2. The control used was referring to the FDA-approved drugs. NNRTIs used as positive controls, while protease inhibitors (PIs) and integrase inhibitor as negative controls [42]. Molecular docking was attempted in triplicate using AutoDock 4.

**Virtual screening of Indonesian herbal database and refinement analysis.** Virtual screening was performed using PyRx software with the best parameters of the control compounds orientation docking. Grid parameter used was $50 \times 50$ units with 0.375 Å per unit. Docking parameter was set at 250,000 calculations; 27,000 generations; 150 populations; mutation and crossover rate at default. Virtual screening was applied 5 times. Top 10

Table 5.2.   Docking result of control compounds using AutoDock with HIV-1 RT as target [1a].

| Rank based on $\Delta$G | Name | $\Delta$G (kcal/mol) (N = 3) | SD (kcal/mol) | CV (%) |
|---|---|---|---|---|
| 1 | Etravirine* | −8.82 | 0.030 | 0.340 |
| 2 | Nevirapine* | −7.85 | 0.010 | 0.127 |
| 3 | Rilpivirine* | −7.61 | 0.221 | 2.897 |
| 4 | Efavirenz* | −7.33 | 0.026 | 0.361 |
| 5 | Delavirdine* | −6.72 | 0.516 | 7.679 |
| 6 | Raltegavir** | −4.49 | 2.328 | 51.887 |
| 7 | Amprenavir | −0.49 | 2.829 | 573.403 |
| 8 | Tipranavir** | 5.37 | 8.278 | 154.146 |
| 9 | Darunavir** | 6.55 | 7.410 | 113.135 |
| 10 | Nelfinavir** | 15.82 | 14.480 | 91.528 |
| 11 | Lopinavir** | 83.82 | 39.737 | 47.406 |
| 12 | Saquinavir | 160.33 | 69.265 | 43.202 |
| 13 | Ritonavir** | 175.55 | 82.990 | 47.273 |
| 14 | Atazanavir** | 294.38 | 69.715 | 23.682 |

*Positive control used for this research was etravirine, nevirapine, rilpivirine, efavirenz, and delavirdine. These are a member of HIV-1 NNRTIs (non-nucleoside reverse transcriptase inhibitors).
**Negative control used for this research were raltegravir (member of integrase inhibitor), amprenavir, tipranavir, darunavir, nelfinavir, lopinavir, saquinavir, ritonavir, and azatanavir (member of protease inhibitors).

compounds screened were refined after docking. The parameter was set at 2,500,000 calculations. The best result then analyzed to observe its binding site and interactions using PyMOL [43] binding site was defined as residues in the proximity of 5 Å from the docked ligand pose.

### 5.3.1.2.   *Results*

#### 5.3.1.2.1.   3D structure of the chemical compounds from medicinal plants in Indonesia

Initially, a total of 1,412 3D structure of chemical compounds from medicinal plants of Indonesia embedded in the system which can be accessed at http://herbaldb.farmasi.ui.ac.id (Fig. 5.2). The website is an open system

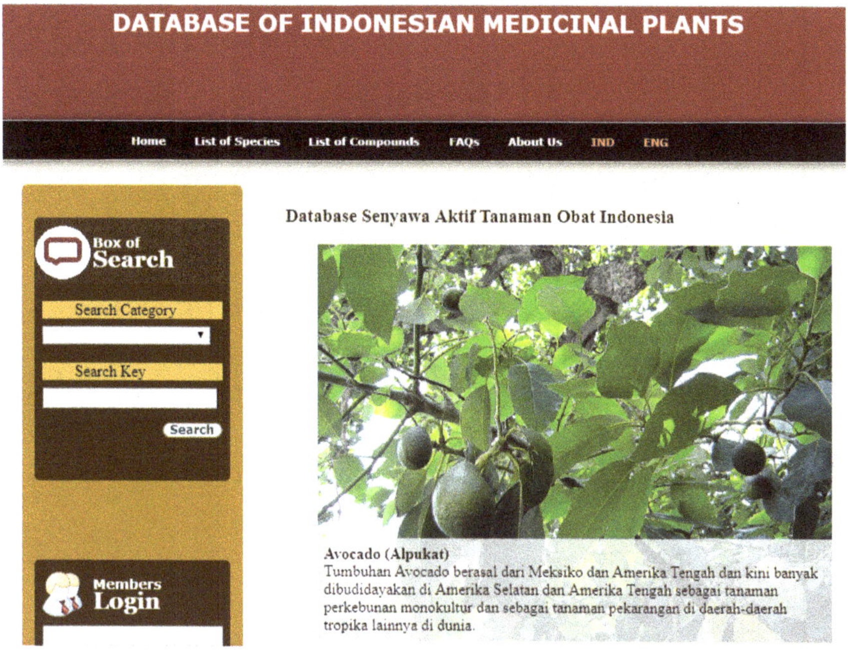
DATABASE OF INDONESIAN MEDICINAL PLANTS
Home
List of Species
List of Compounds
FAQs
About Us
IND
ENG
Box of Search
Search Category
Search Key
Search
Members Login
Database Senyawa Aktif Tanaman Obat Indonesia
Avocado (Alpukat)
Tumbuhan Avocado berasal dari Meksiko dan Amerika Tengah dan kini banyak dibudidayakan di Amerika Selatan dan Amerika Tengah sebagai tanaman perkebunan monokultur dan sebagai tanaman pekarangan di daerah-daerah tropika lainnya di dunia.

Fig. 5.2. Web Database of medicinal plants and its 3D structure.

and allows wide use by the general public and scientists or other stakeholders from industry, government, and university.

**Virtual screening of the Herbal database.** First, the preliminary docking of 14 control compounds was performed. Five compounds as positive controls which were known to have inhibition activity to HIV-1 RT: etravirine, nevirapine, rilpivirine, efavirenz, and delavirdine. Then, the results of docking were ranked based on its best docked binding energy as shown in Table 5.3. Based on the optimized parameter, positive controls outrank all negative controls. With that parameter, virtual screening was applied to the Indonesia herbal database [44].

Top 10 compounds were given in Table 5.2. The average binding energies vary from −11.28 kcal/mol to −10.36 kcal/mol. Moreover, the top three of them linked to the current research shows their inhibitor effect toward HIV-1 activities. The first rank compound, mulberrin, alongside with

Table 5.3.    Top 10 compounds docked by Auto Dock based on its binding energy with HIV-1 RT [1a].

| Rank | Name | $\Delta$G (kcal/mol) (N = 3) | N | SD (kcal/mol) | CV (%) | Plant(s) source [13, 23] |
|---|---|---|---|---|---|---|
| 1 | Mulberrin | −11.28 | 2 | 0.3041 | 2.697 | *Artocarpus fretessi, A. gomezianus Wallich ex Trecul, A. heterophyllus, Morus Alba, M. australis, M. mongolica* |
| 2 | Plucheoside A | −10.82 | 4 | 0.3293 | 3.044 | *Pluchea indica* |
| 3 | Vitexilactone | −10.74 | 5 | 0.0438 | 0.408 | *Vitex cannabifolia, V. cannabinifolia, V. trifolia, Tinospora rumphii* |
| 4 | Brucine N-oxide | −10.70 | 5 | 0.0614 | 0.574 | *Strychnos atlantica, S. lucida R. Br., S. spinose, S. wallichiana* |
| 5 | Cyanidin 3-arabinoside | −10.66 | 4 | 0.1800 | 1.689 | *Mangifera Indica, Acrotriche serrulata, empetrum ningrum, epacris gunni, Gaylussacia spp., Leucopogon collinus, Rhododendron spp., Vaccinium padifolium, Sterculia parviflora, Theobroma cacao, penstemon spp. phalaris arundinacea, Polygonum spp. Aronia melanocarpa, malus sylvestris, Cinchona ledgeriana, Saxifraga spp. Camelia japonica* |
| 6 | Alpha-mangostin | −10.51 | 4 | 0.2130 | 2.028 | *Allanblackia monticola STANER L.C., Garcinia kowa, G. dulcis, G. echinocarpa, G. fusca, G. Mangostana, G. terpnophylla, Cratoxylum cochinchinese* |
| 7 | Guajaverin | −10.49 | 3 | 0.1270 | 1.210 | *Foeniculum vulgare, Arctostaphylos uva-ursi, Calluna vulgaris, Chamaedaphne calyculata, Richea angustifolia, R. scoparia, Hypericum erectum Thunb, Hibiscus mutabilis, Theobroma, cacao L., Eucalyptus cypellocarpa, Psidium guajava, Securidaca diversifolia, polygonium aviculare, Zanthoxylum bungeanum, taxodium distichum* |
| 8 | Erycristagallin | −10.43 | 3 | 0.3402 | 3.261 | *Erythrina abyssinica, E. crista-galli, E. orientalis, E. subumbrans, E. variegate* |
| 9 | Morusin | −10.43 | 5 | 0.3290 | 3.155 | *Artocarpus fretessi, Morus alba, M. australis, M. mongolica* |
| 10 | Sanggenol N | −10.36 | 4 | 0.1668 | 1.611 | *Morus australis* |

morusin and sanggenol N (ranked 9 and 10, respectively) were flavonoids from Moraceae family, notably from *Artocarpus fretessi, A. gomezanius* Wallich ex Trexcul, *A. heterophyllus, Morus alba, M. australis poir*, and *M. mongolica* [44–49]. Recent research shows that flavonoids from *Morus alba* (e.g. morusin, kuvanon H and morusin-4'-glucoside) have anti-HIV activities [50]. The second rank compound, plucheoside A, is a terpeneg-lycoside from *Pluchea indica* [44, 45, 51]. Based on research of Locher *et al.* which has studied the anti-HIV-1 activities of Hawaiian plants. It suggests *Pluchea indica* could inhibit HIV-1 activities, however, the mechanism of inhibition was not revealed further [52]. The third rank compound, vitexilactone is a diterpene from vitex cannabivolia, *V. cannabinifolia, V. trifolia*, and *Tinospora rumphii* [44, 45, 53–55]. This plant is known well as a treatment for HIV-AIDS and shown anti-HIV-1 RT activities [56].

### 5.3.2. *Study II: Molecular Docking and Molecular Dynamics Analysis of Vietnam Herbal to Treat Alzheimer*

In all, 342 compounds derived from Vietnamese plants were collected and studied their binding affinity to full-length $A\beta_{1-40}$ and $A\beta_{1-42}$ peptides and their mature fibrils using the Autodock vina version 1.1 [57] and molecular dynamics (MD) simulations. Top-leads found by the docking technique are further refined by the more accurate molecular Mechanic–Poisson–Boltzmann surface area (MM–PBSA) method. As the results, this study predicts that five ligands: Dracorubin; taraxerol; taraxasterol; hinokiflavone; and diosgenin; are good candidates for treating AD. In designing oral drugs for the AD, it is important to know whether they can be absorbed by the human body and then pass the blood–brain barrier (BBB). Furthermore, in molecular dynamics, the specification and the speed of the computer really matter. The need to improve the speed of long processes used in the analysis of molecular dynamics in order to search the potential drug to treat the diseases. This session also will show how to improve the performance of MD simulations.

## 5.3.2.1. *Material and methods*

### 5.3.2.1.1. Set of receptor-ligand complexes

Four receptors were considered, including monomers $A\beta_{1-40}$ and $A\beta_{1-42}$ and protofibrils of their fragments $6A\beta_{9-40}$ and $5A\beta_{17-42}$ (first 8 and 16 unstructured amino acids re-excluded from mature fibrils). The NMR structures of $A\beta_{1-40}$ (PDB ID: 1BA4 [58]) and $A\beta_{1-42}$ (PDB ID 1Z0Q [59]) peptides were taken from the PDB. Coordinates of twofold symmetry $6A\beta_{9-40}$ were provided by Tycko [60], whereas the crystal structure of $5A\beta_{17-42}$ was taken from the PDB (PDB ID: 2BEG [61]). 1BA4 and 2BEG have 10 models whereas 1Z0Q has 30 models. Their first model was chosen for the docking simulation. On the other hand, this study collected 342 compounds derived from Vietnamese herbs [62] for the ligands, with their chemical structures are known from the PubChem and ChemSpider database (http://pubchem.ncbi.nlm.nih.gov and http://www.chemspider.com/ respectively). The general assisted model building with energy refinement (AMBER) force field [28] had been used to generate ligand parameters, except for charges that came from molecular orbital PACkage (MOPAC) using the Austin model 1-bond charge correction (AM1-BCC) [63]. This procedure was done by Ambertools-1.4.

**Molecular docking simulation.** This study prepared PDBQT file for receptor and ligand using AutodockTools 1.5.4 in order to perform docking of ligands to full-length $A\beta$ peptides and their fibrils [64]. Atomic interactions were described by a modified version of the Chemistry at Harvard Molecular Mechanics (CHARMM) force field [65, 66]. The Broyden–Fletcher–Goldfarb–Shanno method [67] was implemented for local optimization. For the optimal result, the exhaustiveness of global search was set equal to 400 whereas the maximum energy difference between the best and worst binding modes was chosen to be 7. Twenty modes of docking were generated with random starting positions of ligand that enabled the fully flexible torsion degrees of freedom. Positions of all atoms of the receptor were kept fixed. The grid center was placed at the center of mass of receptor and grid dimensions were set large enough ($60 \times 50 \times 50$, $70 \times 50 \times 50$, $90 \times 70 \times 50$ and $80 \times 60 \times 60$ Å for $A\beta40$, $A\beta42$, $6A\beta9$–$40$ and $5A\beta17$–$42$, respectively) to cover every part of the receptor.

**The blood–brain barrier (BBB) and Human intestinal absorption (HIA).** Since amyloid peptides are located in the brain, an efficient drug should be able to cross the BBB to interfere with their activity. The ability to cross BBB was measured by the logarithm base 10 of the ratio of the compound concentration in the brain ($C_{\text{brain}, \_}$ to that in the blood ($C_{\text{blood}}$) as shown in Eq. (5.1). Another important aspect of the oral drug design is HIA [68] the percentage of drug that can be absorbed by the human body. HIA should be high enough for drug efficacy. According to 'rule of 5' [69, 70], it depends on molecular weight, number of HB donors, number of HB acceptors, C log P, and M log P. HIA of all compounds was estimated by the QSAR method [25, 38, 40, 41] which was also implemented in the PreADME suit [69].

$$\log(BB) = \log\left[\frac{C_{\text{brain}}}{C_{\text{blood}}}\right] \tag{5.1}$$

**Molecular dynamics simulations.** AMBER 10 package [70] was utilized to run MD simulations with the AMBER 99SB force field. The water model TIP3P [71] was chosen following the recommendation of this package. Nine complexes of $6A\beta_{9-40}$-ligands were placed in a triclinic box of about 9500 water molecules with 0.7 nm distance between the solute and box. $6A\beta_{9-40}$ has 2862 atoms, whereas Dracorubin, Solasodine, Taraxasterol, Amentoflavone, Hinokiflavone, Kulolactone, Hecogenin, Taraxerol, and Diosgenin have 61, 74, 81, 58, 58, 79, 73, 81, and 72 atoms, respectively. Furthermore, to neutralize the systems, six $Na^+$ ions were added, except for the complex with ligand Solasodine into which five $Na^+$ ions were added. The long-range electrostatic interaction is computed by particle-mesh Ewald summation method [72]. Equations of motion were integrated using a leap-frog algorithm [73] with a time step of 2 fs. The non-bonded interaction pair-list was updated every 10 fs with the cutoff of 0.8 nm. The systems were minimized to remove bad vdW contacts with water. Then the temperature was gradually increased from 0 to 300 K for 50 ps. For density equilibration, MD simulation was carried out with weak restraints on all bonds of the complex for 50 ps at a constant temperature of 300 K [74, 75]. Restraints have been implemented by the linear constraint solver (LINCS) algorithm [76]. Constant temperature 300 K was enforced using Berendsen algorithm [77] under 500 ps canonical ensemble (NVT) simulation with a damping coefficient of 0.1 ps. The final MD simulation was carried out in the isothermal–isobaric (NPT) ensemble using Parrinello–Rahman pressure coupling [78] at 1 atm with the damping coefficient of 0.5 ps.

**Binding free energy calculation by MM–PBSA.** The binding free energy is defined by this following equation:

$$\Delta G_{bind} = G_{complex} - G_{free-protein} - G_{free-ligand} \tag{5.2}$$

In the MM–PBSA, the free energy of each molecule is given by the following equation:

$$G = E_{mm} - G_{Solvation} - TS \tag{5.3}$$

The molecular mechanic's energy of the solute in the gas phase $\mathbf{E}_{mm}$ includes bond, bond-angle, dihedral-angle, electrostatic, and vdW (Lennard–Jones) terms:

$$E_{mm} = E_{bond} + E_{angle} + E_{torsion} + E_{elec} + E_{vdW} \tag{5.4}$$

To incorporate all possible non-bonded interactions, $\mathbf{E}_{mm}$ was computed without cut-off utilizing AMBER tool. The free energy of solvation, $\mathbf{G}_{solvation}$, was approximated as the sum of electrostatic and non-polar contributions,

$$G_{Solvation} = G_{PB} + G_{sur} \tag{5.5}$$

$\mathbf{G}_{PB}$ derived from the electrostatic potential between solute and solvent was determined using the continuum solvent approximation [51] from a low solute dielectric constant region ($\varepsilon = 2$) to higher one with water without salt ($\varepsilon = 78.45$). Using grid spacing 0.1 Å, the APBS package [51] was implemented for the numerical solution of the corresponding linear Poisson–Boltzmann equation. The GROMOS radii and charges were used to generate PQR files. Then the non-polar solvation term $\mathbf{G}_{sur}$ was approximated as linearly dependent on the solvent-accessible surface area (SASA), derived from Shrake–Rupley numerical method [79] integrated into the

APBS package. $G_{sur} = \gamma SASA + \beta$, where $\gamma = 0.0072 \, kcal/mol \, \text{Å}$ and $\beta = 0$ [80].

Solute entropy contributions were estimated from the structures obtained in the equilibrium. In the MM–PBSA method, the conformational entropy of the solute is approximated by the vibrational entropy $S_{vib}$ that is estimated from normal mode analysis by diagonalizing the mass-weighted Hessian matrix [79] as follows:

$$S_{vib} = -R \ln(1 - e^{-hv_0/K_bT}) + \frac{N_A V_0 e^{-hv_0/KBT}}{T(1 - e^{-hv_0/KBT})} \tag{5.6}$$

where $h$ is Plank's constant, $v_0$ is the frequency of the normal mode, $K_B$ is the Boltzmann constant, $T = 300 \, K$, and $N_A$ is Avogadro's number. Then the snapshots were used to collect every 10 ps in the equilibrium to compute other terms of $\Delta G_{bind}$.

### 5.3.2.1.2. Big Data software to run MD simulation

Most of MD simulations are time-consuming processes, therefore, it needs high-performance computing infrastructures to run the simulations. For example, there were researchers who used a single GTX GPU which provided improved performance compared to GPUs in a cluster environment. Researchers have also used AMBER to accelerate the computation time of molecular dynamics. It shows that the computational time is faster, 25–33 times than the sequential process [81]. Therefore, each time to process the molecular dynamics simulation, we need to consider the speed of the simulation.

### 5.3.2.2. *Results*

### 5.3.2.2.1. Human intestinal absorption

This study calculates HIA for 342 compounds. This value varies between 0% and 100%, with an average value of 81%, implying that most of the ligands can be absorbed by the human body. Among them, 50 compounds have 100% absorption, and 227 compounds have HIA of 90%. Only six ligands are not able to penetrate the body (0% HIA). Curcumin, which is a potential drug for treating AD, has high HIA of 94%, whereas other candidates have relatively low absorption ability. For instance, HIA = 65%, 40%, 21%, 40%, and 21% for Ginkgolide A, Ginkgolide B, Ginkgolide C, Ginkgolide J, and EGCG, respectively. Thus, most of the ligands display higher absorption ability than leads compounds that are under intensive clinical trial.

### 5.3.2.2.2.   Blood–brain barrier

Using the PreADME prediction method, log (BB) were calculated for the ligands (Eq. (1)), which measures a percentage of drug that can permeate the brain. Experimental values of log (BB) of drugs published to date cover the range between $-2.0$ and $+1.0$ [68]. Compounds with log (BB). $>0.3$ can cross the BBB easily, whereas compounds with log (BB), $<-1.0$ are poorly distributed in the brain [69]. The average value of log (BB) is $-0.31$. Compound Taraxasterol has the largest penetration ability log (BB) = 1.36, whereas ligand 10,607 has the smallest penetration ability log (BB) = $-2.0$. 91 compounds are observed with log (BB) around $<-1.0$ and 24 compounds that have log (BB) larger than 1.00. At least 227 compounds can pass through the BBB easily. There is a weak correlation between HIA and log (BB) with the correlation level R = 0.54.

### 5.3.2.2.3.   Molecular docking results

Autodock Vina version 1.1 [57] was used to carry out docking of 342 ligands to $A\beta_{1-40}$, $6A\beta_{9-40}$, $A\beta_{1-42}$ and $5A\beta_{17-42}$. The distributions of $\mathbf{E}_{\text{bind}}$ for four sets are shown in Fig. 5.3. Two sets of binding energies to fibril $6A\beta9$–40 and monomer $A\beta_{1-40}$ display high correlation with the correlation level R = 0.91 (Fig. 5.4). In the case of the longer 42-bead peptide, the correlation level drops to $\mathbf{R} = 0.78$ for targets $A\beta_{1-42}$ and $5A\beta_{17-42}$. For the remaining four pairs [$6A\beta_{9-40}$, $5A\beta_{17-42}$], [$6A\beta_{9-40}$, $A\beta_{1-42}$], [$A\beta_{1-40}$, $5A\beta_{17-42}$], and [$A\beta_{1-40}$, $A\beta_{1-42}$], the correlation level is $\mathbf{R} = 0.80, 0.94, 0.74$, and $0.92$, respectively. Thus, $\mathbf{E}_{\text{bind}}$ obtained for $A\beta_{1-42}$ shows the highest correlation with $6A\beta_{9-40}$ ($\mathbf{R} = 0.94$) but not with $5A\beta_{17-42}$.

Overall, the correlation between the four sets of $\mathbf{E}_{\text{bind}}$ is high, but this does not mean they provide the same binding affinity ranking. Solasodine and Diosgenin show the highest susceptibility to $6A\beta_{9-40}$ with

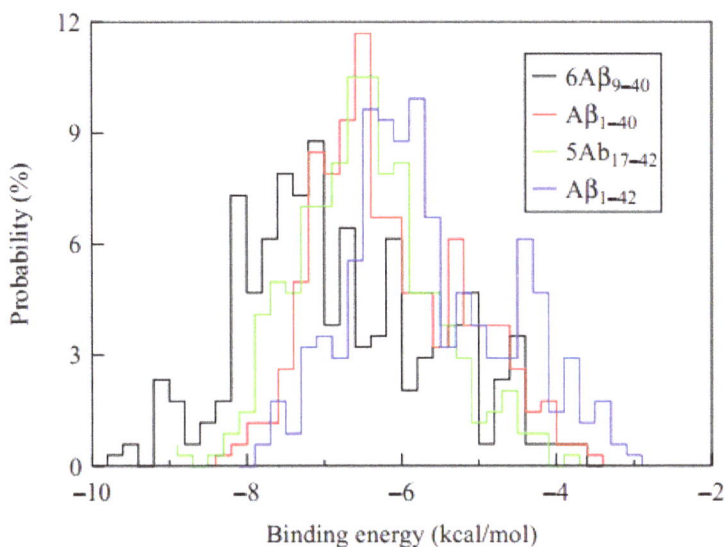

Fig. 5.3.  Distributions of binding energies of 342 ligands to four receptors [25].

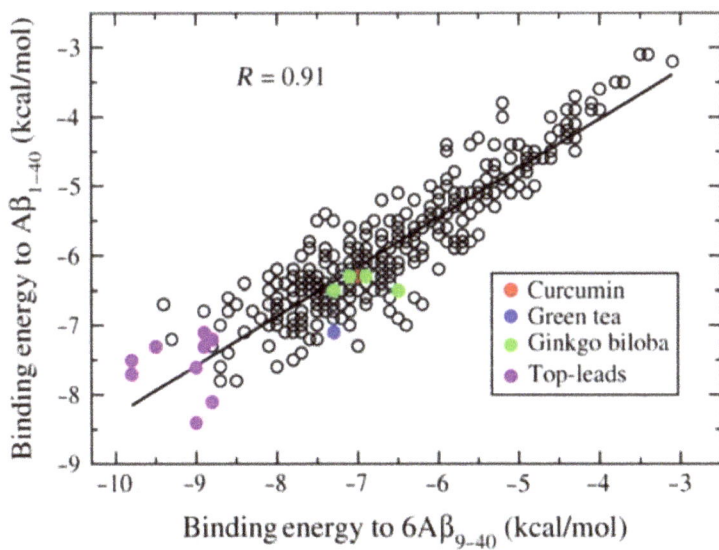

Fig. 5.4.  Relationship between binding energies to Ab1-40 and 6AB9-40 [25].

$E_{bind} = -9.8$ kcal/cal, whereas Kulolactone has the lowest binding energy $E_{bind} = -8.9$ kcal/cal to $5A\beta_{17-42}$. If we make ranking by binding energies to $A\beta_{1-40}$, then Dracorubin occupies the first place having $E_{bind} = -8.4$ kcal/cal. Sorting ligands by $E_{bind}$ to monomer $A\beta_{1-42}$, Amentoflavone occupies the first place with $E_{bind} = -8.1$ kcal/cal. It should be noted that binding energies are not correlated with either log (BB) or HIA.

**Molecular dynamics simulation.**   Since the docking method is not accurate enough for prediction, we refine our finding by calculating the binding free energy of nine top-leads using a more reliable MM–PBSA method. Since the results of docking study obtained for different targets show high correlation, only $6A\beta_{9-40}$ was chosen as a receptor for MD simulations. For each system, we had carried out MD run of 19 ns except that MD run of 29 ns simulation was carried out for compound Solasodine. Since all systems behave similarly, we show results for four ligands, Dracorubin, Hinokiflavone, Kulolactone, and Hecogenin, that have very different binding free energies (see Table 5.2). From the time dependence of backbone RMSD from the initial structures (Fig. 5.5), it is clear that these systems get equilibrium at different time $t_{eq}$ ($t_{eq} < 13, 9, 13$, and 11 ns for Dracorubin, Hinokiflavone, Kulolactone, and Hecogenin, respectively).

The ranking of binding affinity obtained by the MMPBSA method (Table 5.4) is very different from the docking ranking as the correlation level between two sets of results is almost zero. Ligand 160270 ranked fourth by docking (Table 5.5) becomes the first place in MM–PBSA (Table 5.2). Since the latter method is more accurate, we should rely on its results. Keeping only ligands that have $\Delta G_{bind}, < -11$ kcal/mol, we predict that Dracorubin, Taraxerol, Taraxasterol, Hinokiflavone, and Diosgenin may be good candidates to cope with AD. Using the relationship between the binding free energy and inhibition constant $K_i$ ($\Delta G_{bind} = RT \ln(Ki)$), where the gas constant $R = 1.987 \times 10^{-3}$ kcal/mol), we can show that $K_i$ of five top-leads varies between 8 nM and 4 pM. In other words, they have the excellent inhibitory capacity. Having used the MM–PBSA and the same force field and water model, we have obtained the binding free energy of Curcumin to $6A\beta_{9-40} \Delta G_{bind} \approx -14.3$ kcal/mol (Son Tung Ngo and Mai Suan Li,

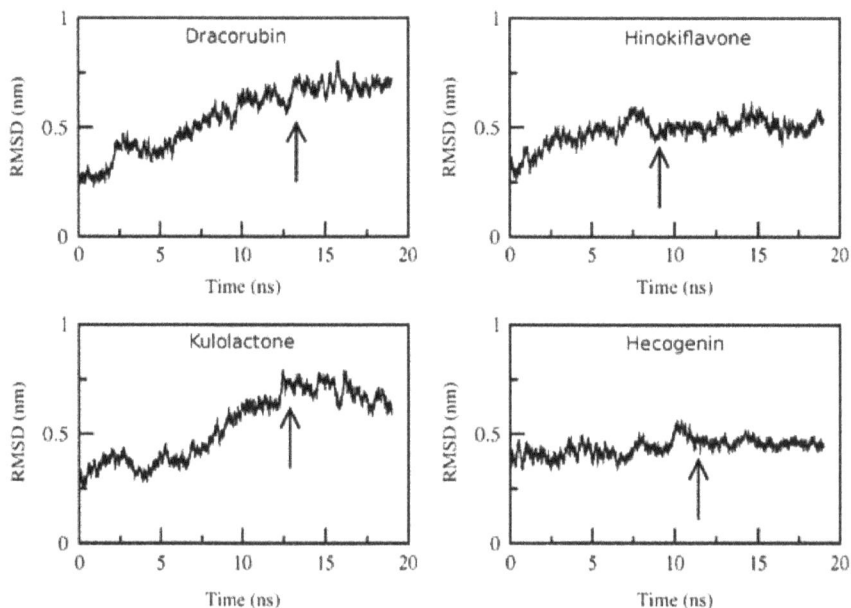

Fig. 5.5. Time dependence of RMSD for four compounds. Arrow refers to equilibrium time $t_{eq}$ when the system reaches equilibrium.

Table 5.4. Binding free energies to 6A$\beta_{9-40}$ of nine top-leads revealed by docking method [25].

| ID | Ligand | $-T\Delta S$ | $\Delta E_{vdw}$ | $\Delta E_{elec}$ | $\Delta G_{PB}$ | $\Delta E_{sur}$ | $\Delta E_{bind}$ |
|---|---|---|---|---|---|---|---|
| 160270 | Dracorubin | 21.22 | −58.36 | −12.06 | 37.73 | −4.12 | −15.59 |
| 92097 | Taraxerol | 19.04 | −41.73 | −1.07 | 12.76 | −3.85 | −14.85 |
| 5270604 | Taraxasterol | 20.25 | −50.88 | 0.03 | 22.42 | −3.95 | −12.13 |
| 5281627 | Hinokiflavone | 24.16 | −55.69 | −18.34 | 42.41 | −4.39 | −11.85 |
| 99474 | Diosgenin | 20.36 | −52.25 | −4.35 | 29.61 | −4.47 | −11.1 |
| 91453 | Hecogenin | 19.69 | −50.32 | −3.07 | 31.77 | −4.21 | −6.14 |
| 5281600 | Amentoflavone | 21.91 | −43.57 | 0.26 | 19.88 | −3.26 | −4.88 |
| 31342 | Solasodine | 23.25 | −38.60 | −118.29 | 132.4 | −3.23 | −4.37 |
| 5318868 | Kulolactone | 22.68 | −45.03 | 7.39 | 16.86 | −4.30 | −2.40 |

Table 5.5.   Nine top-leads revealed by ranking binding energies [25].

| ID | Herbs name | Compound | $6A\beta_{9-40}$ | $A\beta_{1-40}$ | $5A\beta_{17-42}$ | $A\beta_{1-42}$ | Log (BB) | HIA (%) |
|---|---|---|---|---|---|---|---|---|
| 31342 | *S. xantho-carpum Schrad* | Solasodine | −9.8 | −7.5 | −7.5 | −7.6 | 0.84 | 94 |
| 99474 | *Schizocapsa plantaginea Hance* | Diosgenin | −9.8 | −7.7 | −7.4 | −7.5 | 0.89 | 96 |
| 5281627 | *T. orientalis L* | Hinokiflavone | −9.5 | −7.3 | −8.0 | −7.6 | −0.94 | 87 |
| 160270 | *C. draco Willd* | Dracorubin | −9.0 | −8.4 | −8.7 | −7.3 | −1.25 | 98 |
| 5270604 | *Centipeda Minima* | Taraxasterol | −9.0 | −7.6 | −7.5 | −7.0 | 1.36 | 100 |
| 5318868 | *Melia Azedarach L* | Kulolactone | −8.9 | −7.1 | −8.9 | −7.1 | 1.03 | 96 |
| 91453 | *Agave American Lin* | Hecogenin | −8.9 | −7.3 | −7.6 | −7.4 | 0.26 | 96 |
| 92097 | *T.offcinale* | Taraxerol | −8.8 | −7.2 | −7.9 | −7.2 | 1.32 | 100 |
| 5281600 | *Selaginella tamariscina* | Amentoflavone | −8.8 | −8.1 | −7.5 | −8.1 | −0.93 | 81 |

unpublished results). Using the experimental value $Ki = 0:2\,nM$ [55], we obtain $\Delta G_{bind} \approx -13.3\,kcal/mol$ for binding of Curcumin to $A\beta_{9-40}$ aggregates. Therefore, the binding affinity of Dracorubin and Taraxerol to the $A\beta$ fibrils is probably compatible or even higher than Curcumin.

### 5.3.3. *Study III: Transcriptomic Role as a Tool in Early-stage Drug Discovery Decision Making*

In the developing of drugs, failures in the late stage are extremely costly because large amounts of time and capital have already been invested. Gene expression data were informative and able to support yes and no decisions. This discussion show that gene expression profiling can detect adverse effects of compounds and is a valuable tool in early-stage drug discovery decision making.

#### 5.3.3.1. *Material and methods*

The mRNA expression data were quantile normalized, summarized [57], and filtered [82, 83]. Subsequent exploratory analysis to detect strong transcriptomic effects was performed using spectral map analysis [84]. Differentially expressed genes [85] were called, and transcriptional modules [86] (i.e. genes where expression is simultaneously up- or down-regulated in a subset of samples) were identified by the factor analysis for bicluster acquisition (FABIA) method [87]. Transcriptional modules related to the desired effect were identified by the potential support vector machine (PSVM) using target-related bioassay measurements [88]. A data framework and analysis pipeline were constructed to facilitate an integrated analysis of gene expression data, chemical structures, and bioassay results.

Table 5.6. Overview of the pharmaceutical projects included in the QSTAR project for which transcriptomic profiling was performed [23].

| Target | Therapeutic area | Result | Utility | Decision |
|---|---|---|---|---|
| ROS1 | Oncology | Selectivity and on-target | Useful | No-go for certain chemotypes |
| EGFR | Oncology | On-target and off-target | Useful | No-go and go for certain compounds |
| PDE10A | Neuroscience | Off-target | Useful | No-go for certain compounds |
| MTP | Metabolic | On-target (inconsistency with assay data) | Relevant | |
| MGluR2PAM | Neuroscience | Off-target (further exploration needed) | Relevant | |
| FGFR | Oncology | On-target (no differentiation among compounds) | Relevant | |
| HBV | Virology | Limited GE effects | No added value | |
| PDE4 | Neuroscience | Limited GE effects | No added value | |

## 5.3.3.2.  *Results*

Table 5.6 provides an overview of the projects and disease areas that were explored in this study about the utility of using transcriptomic data (measured by microarrays) in the lead optimization phase of pharmaceutical drug discovery projects. Transcriptomic data provided relevant information for six of the eight projects. For three of them (ROS1, EGFR, and PDE10A), the data provided clear go or no-go decisions. In three other projects (FGFR, mGluR2PAM, and MTP), transcriptomic delivered novel biological insights but did not provide direct decision support. In the remaining two cases (HBV and PDE4), neither biological insights nor go or no-go decisions were gained.

## 5.4.  Conclusion

*In silico* study is very useful for the development of drugs. The combination of some methods such as molecular docking, molecular dynamics, and gene expression data analysis is really helpful to create a good and valid drug prediction through dry-lab analysis. Natural compounds are an alternative compound which can be used as a treatment for AIDS and DA since the current treatment show ineffective in combating that two diseases. In study 1, it shows a database of natural products in Indonesia and the analysis of molecular docking to gain a novel treatment from that natural product toward AIDS. Furthermore, in the second study, it shows the herb product from Vietnam can be an alternative option to treat AD patients. Molecular docking and molecular dynamics were performed. In study 3, it shows gene expression data were informative and able to support go or no-go decisions. The combination of those three methods from study one, two, and three can yield a better and more valid prediction.

## References

1.   1. (a) Stephens, Z.D., Lee, S.Y., Faghri, F., Campbell, R.H., Zhai, C., Efron, M.J., Iyer, R., Schatz, M.C., Sinha, S., Robinson, G.E. (2015). Big data: astronomical or genomical? *PLoS biology* 13(7), e1002195. (b) Syahdi, R. R., Muñim, A., Suhartanto, H. and Yanuar, A. (2012). Virtual screening of Indonesian herbal database as HIV-1 reverse transcriptase inhibitor. *Bioinformation* 8(24), p. 1206.

2. http://www.prb.org/pdf06/howhivaidsaffectspopulations.pdf.

3. Agholi, M. *et al.* (2012). AIDS Res Hum Retrovirus. PMID: 22873400.

4. http://www.unaids.org/en/media/unaids/contentassets/documents/unaidspublication/2011/JC2216_WorldAIDSday_report_2011_en.pdf.

5. Hardy, J. and Selkoe, D. J. (2002). The amyloid hypothesis of Alzheimer's disease: progress and problems on the road to therapeutics. *Science* 297, pp. 353–356.

6. Scotti, L. and Scotti, M. T. (2018). *In Silico* Studies Applied to Natural Products with Potential Activity Against Alzheimer's Disease. In *Computational Modeling of Drugs Against Alzheimer's Disease* (pp. 513–531). Humana Press, New York, NY.

7. Keputusan Menteri Kesehatan Republik Indonesia No: 381/MenKes/SK/III/2007, Menteri Kesehatan RI, Jakarta, 2007.

8. Jayaram, B. (2011). SCFBIO: what is drug design? http://www.scfbio-iitd.res.in/tutorial/drugdiscovery.htm.

9. Cowlrick, I., Hedner, T., Wolf, R., Olausson, M. and Klofsten, M. (2011). Decision-making in the pharmaceutical industry: analysis of entrepreneurial risk and attitude using uncertain information. *R&D Management* 41(4), pp. 321–336.

10. Gohlmann, H. and Talloen, W. (2009). *Gene Expression Studies Using Affymetrix Microarrays.* CRC Press.

11. Hochreiter, S., Clevert, D. A. and Obermayer, K. (2006). A new summarization method for Affymetrix probe level data. *Bioinformatics* 22(8), pp. 943–949.

12. Schena, M., Shalon, D., Davis, R. W. and Brown, P. O. (1995). Quantitative monitoring of gene expression patterns with a complementary DNA microarray. *Science* 270(5235), pp. 467–470.

13. Lamb, J., Crawford, E. D., Peck, D., Modell, J. W., Blat, I. C., Wrobel, M. J., Lerner, J., Brunet, J. P., Subramanian, A., Ross, K. N. and Reich, M. (2006). The Connectivity Map: using gene-expression signatures to connect small molecules, genes, and disease. *Science* 313(5795), pp. 1929–1935.

14. Bai, J. P., Alekseyenko, A. V., Statnikov, A., Wang, I. M. and Wong, P.H. (2013). Strategic applications of gene expression: from drug discovery/development to bedside. *The AAPS Journal* 15(2), pp. 427–437.

15. van der Veen, J. W., Pronk, T. E., van Loveren, H. and Ezendam, J. (2013). Applicability of a keratinocyte gene signature to predict skin sensitizing potential. *Toxicology in Vitro* 27(1), pp. 314–322.

16. Magkoufopoulou, C., Claessen, S. M. H., Tsamou, M., Jennen, D. G. J., Kleinjans, J. C. S. and van Delft, J. H. M. (2012). A transcriptomics-based *in vitro* assay for predicting chemical genotoxicity in vivo. *Carcinogenesis* 33(7), pp. 1421–1429.

17. Jiang, Y., Gerhold, D. L., Holder, D. J., Figueroa, D. J., Bailey, W. J., Guan, P., Skopek, T. R., Sistare, F. D. and Sina, J. F. (2007). Diagnosis of drug-induced renal tubular toxicity using global gene expression profiles. *Journal of Translational Medicine* 5(1), p. 47.

18. Iorio, F., Bosotti, R., Scacheri, E., Belcastro, V., Mithbaokar, P., Ferriero, R., Murino, L., Tagliaferri, R., Brunetti-Pierri, N., Isacchi, A. and di Bernardo, D. (2010). Discovery of drug mode of action and drug repositioning from transcriptional responses. *Proceedings of the National Academy of Sciences* 107(33), pp. 14621–14626.

19.   Dudley, J. T., Sirota, M., Shenoy, M., Pai, R. K., Roedder, S., Chiang, A. P., Morgan, A. A., Sarwal, M. M., Pasricha, P. J. and Butte, A. J. (2011). Computational repositioning of the anticonvulsant topiramate for inflammatory bowel disease. *Science Translational Medicine* 3(96), pp. 96ra76–96ra76.
20.   Sirota, M., Dudley, J. T., Kim, J., Chiang, A. P., Morgan, A. A., Sweet-Cordero, A., Sage, J. and Butte, A. J. (2011). Discovery and preclinical validation of drug indications using compendia of public gene expression data. *Science Translational Medicine* 3(96), pp. 96ra77–96ra77.
21.   Pacini, C., Iorio, F., Gonçalves, E., Iskar, M., Klabunde, T., Bork, P. and Saez-Rodriguez, J. (2012). DvD: An R/Cytoscape pipeline for drug repurposing using public repositories of gene expression data. *Bioinformatics* 29(1), pp. 132–134.
22.   Bol, D. and Ebner, R. (2006). Gene expression profiling in the discovery, optimization and development of novel drugs: One universal screening platform.
23.   Verbist, B., Klambauer, G., Vervoort, L., Talloen, W., Shkedy, Z., Thas, O., Bender, A., Göhlmann, H. W., Hochreiter, S. and QSTAR Consortium (2015). Using transcriptomics to guide lead optimization in drug discovery projects: Lessons learned from the QSTAR project. *Drug Discovery Today* 20(5), pp. 505–513.
24.   Yanuar, A., Mun'im, A., Lagho, A. B. A., Syahdi, R. R., Rahmat, M. and Suhartanto, H. (2011). Medicinal plants database and three dimensional structure of the chemical compounds from medicinal plants in Indonesia. *arXiv preprint arXiv: 1111.7183.*
25.   Ngo, S. T. and Li, M. S. (2013). Top-leads from natural products for treatment of Alzheimer's disease: docking and molecular dynamics study. *Molecular Simulation* 39(4), pp. 279–291.
26.   Heyne, K. (1950). *De Nuttinge Planten van Indonesie*, 3rd edition, Wageningen, H. Veenman & Zonen.
27.   Departemen Kesehatan Republik Indonesia, Materia Medika Indonesia, Vol I. Jakarta: Departemen Kesehatan Republik Indonesia, 1977.
28.   Departemen Kesehatan Republik Indonesia, Materia Medika Indonesia, Vol II. Jakarta: Departemen Kesehatan Republik Indonesia, 1978.
29.   Departemen Kesehatan Republik Indonesia, Materia Medika Indonesia, Vol III. Jakarta: Departemen Kesehatan Republik Indonesia, 1979.
30.   Departemen Kesehatan Republik Indonesia, Materia Medika Indonesia, Vol IV. Jakarta: Departemen Kesehatan Republik Indonesia, 1980.
31.   Departemen Kesehatan Republik Indonesia, Materia Medika Indonesia, Vol V. Jakarta: Departemen Kesehatan Republik Indonesia, 1989.
32.   Departemen Kesehatan Republik Indonesia, Materia Medika Indonesia, Vol VI. Jakarta: Departemen Kesehatan Republik Indonesia, 1995.
33.   Medicinal Herb Index in Indonesia 2nd edition. Jakarta: PT. Eisai Indonesia, 1995.
34.   Bolton, E. E., Wang, Y., Thiessen, P. A. and Bryant, S. H. (2008). PubChem: integrated platform of small molecules and biological activities. In *Annual Reports in Computational Chemistry* (Vol. 4, pp. 217–241). Elsevier.
35.   Nakamura, Y., A Comprehensive Species-Metabolite Relationship Database (KNApSAcK). http://kanaya.naist.jp/KNApSAcK/.
36.   Symyx Draw-An introductory guide, http://bbruner.org/obc/symyx.htm, 2011.

37. Delano, W. (2004). Pymol user's guide. Delano Scientific LLC. http://pymol. sourceforge.net/newman/userman.
38. Guha, R., Howard, M. T., Hutchison, G. R., Murray-Rust, P., Rzepa, H., Steinbeck, C., Wegner, J. and Willighagen, E. L. (2006). The Blue Obelisk — interoperability in chemical informatics. *Journal of Chemical Information and Modeling* 46(3), pp. 991–998.
39. Kusumoto, I. T., Nakabayashi, T., Kida, H., Miyashiro, H., Hattori, M., Namba, T. and Shimotohno, K. (1995). Screening of various plant extracts used in ayurvedic medicine for inhibitory effects on human immunodeficiency virus type 1 (HIV-1) protease. *Phytotherapy Research* 9(3), pp. 180–184.
40. Schliemann, W., Cai, Y., Degenkolb, T., Schmidt, J. and Corke, H. (2001). Betalains of Celosia argentea. *Phytochemistry* 58(1), pp. 159–165.
41. Engvild, K. C. (1986). Chlorine-containing natural compounds in higher plants. *Phytochemistry* 25(4), pp. 781–791.
42. Xiao, D., Kuroyanagi, M., Itani, T., Matsuura, H., Udayama, M., Murakami, M., Umehara, K. and Kawahara, N. (2001). Studies on constituents from Chamaecyparis pisifera and antibacterial activity of diterpenes. *Chemical and Pharmaceutical Bulletin* 49(11), pp. 1479–1481.
43. Lansky, E. P. and Newman, R. A. (2007). Punica granatum (pomegranate) and its potential for prevention and treatment of inflammation and cancer. *Journal of Ethnopharmacology* 109(2), pp. 177–206.
44. Huang, N., Shoichet, B. K. and Irwin, J. J. (2006). Benchmarking sets for molecular docking. *Journal of Medicinal Chemistry* 49(23), pp. 6789–6801.
45. Morikawa, K., Zhao, Z., Date, T., Miyamoto, M., Murayama, A., Akazawa, D., Tanabe, J., Sone, S. and Wakita, T. (2007). The roles of CD81 and glycosaminoglycans in the adsorption and uptake of infectious HCV particles. *Journal of Medical Virology* 79(6), pp. 714–723.
46. Choudhary, M. I., Begum, A., Abbaskhan, A., Musharraf, S. G. and Ejaz, A. (2008). Two new antioxidant phenylpropanoids from Lindelofia stylosa. *Chemistry & Biodiversity* 5(12), pp. 2676–2683.
47. Fossen, T. and Andersen, Ø. M. (2003). Anthocyanins from red onion, Allium cepa, with novel aglycone. *Phytochemistry* 62(8), pp. 1217–1220.
48. Pascale, Steven Patrick. Note pad and organizer clip. U.S. Patent Application 29/320,755, filed February 10, 2009.
49. Jensen, S. R., Franzyk, H. and Wallander, E. (2002). Chemotaxonomy of the Oleaceae: iridoids as taxonomic markers. *Phytochemistry* 60(3), pp. 213–231.
50. Lee-Huang, S., Zhang, L., Huang, P. L., Chang, Y. T. and Huang, P. L. (2003). Anti-HIV activity of olive leaf extract (OLE) and modulation of host cell gene expression by HIV-1 infection and OLE treatment. *Biochemical and Biophysical Research Communications* 307(4), pp. 1029–1037.
51. Knapp, Furn F., John Goad, L. and Trevor, Goodwin, W. (1977). Stereochemistry of C-4 demethylation during conversion of obtusifoliol into poriferasterol by Ochromonas malhamensis. *Phytochemistry* 16(11), pp. 1677–1681.
52. Begum, S., Hassan, S. I., Siddiqui, B. S., Shaheen, F., Ghayur, M. N. and Gilani, A. H. (2002). Triterpenoids from the leaves of Psidium guajava. *Phytochemistry* 61(4), pp. 399–403.

53. Abdel Bar, F. M., Zaghloul, A. M., Bachawal, S. V., Sylvester, P. W., Ahmad, K. F. and El Sayed, K. A. (2008). Antiproliferative triterpenes from Melaleuca ericifolia. *Journal of Natural Products* 71(10), pp. 1787–1790.
54. Lee, C.-K. (1998). A new norlupene from the leaves of Melaleuca leucadendron. *Journal of Natural Products* 61(3), pp. 375–376.
55. Ruzicka, L. and Rosenkranz, G. (1940). To the knowledge of the triterpenes. (54th Communication). About Lupenal and Lupenalol, as well as their further transformations. *Helvetica Chimica Acta* 23(1), pp. 1311–1324.
56. Fujioka, T., Kashiwada, Y., Kilkuskie, R. E., Cosentino, L. M., Ballas, L. M., Jiang, J. B., Janzen, W. P., Chen, I. S. and Lee, K. H. (1994). Anti-AIDS agents, 11. Betulinic acid and platanic acid as anti-HIV principles from Syzigium claviflorum, and the anti-HIV activity of structurally related triterpenoids. *Journal of Natural Products* 57(2), pp. 243–247.
57. Vina, A. (2010). Improving the speed and accuracy of docking with a new scoring function, efficient optimization, and multithreading Trott, Oleg; Olson, Arthur J. *Journal of Computational Chemistry* 31(2), pp. 455–461.
58. Coles, M., Bicknell, W., Watson, A. A., Fairlie, D. P. and Craik, D. J. (1998). Solution structure of amyloid $\beta$-peptide (1–40) in a Water–Micelle environment. Is the membrane-spanning domain where we think it is? *Biochemistry* 37(31), pp. 11064–11077.
59. Tomaselli, S., Esposito, V., Vangone, P., Nulandvan, N. A. Bonvin, A. M., Guerrini, R., Tancredi, T., Temussi, P. A. and Picone, D. (2006). The alpha-to-beta conformational transition of Alzheimer's Abeta-(1-42) peptide in aqueous media is reversible: A step by step conformational analysis suggests the location of beta conformation seeding. *ChemBiochem* 7, pp. 257–267.
60. Petkova, A. T., Yau, W. M. and Tycko, R. (2006). Experimental constraints on quaternary structure in Alzheimer's $\beta$-amyloid fibrils. *Biochemistry* 45(2), pp. 498–512.
61. Lührs, T., Ritter, C., Adrian, M., Riek-Loher, D., Bohrmann, B., Döbeli, H., Schubert, D. and Riek, R. (2005). 3D structure of Alzheimer's amyloid-$\beta$ (1–42) fibrils. *Proceedings of the National Academy of Sciences of the United States of America* 102(48), pp. 17342–17347.
62. Loi, D. T. (2004). *Vietnam's Herbal Plants and Remedies e$^{-}$ (in Vietnamese)*, Medicine Publisher, Ha Noi.
63. Jakalian, A., Jack, D. B. and Bayly, C. I. (2002). Fast, efficient generation of high-quality atomic charges. AM1-BCC model: II. Parameterization and validation. *Journal of Computational Chemistry* 23(16), pp. 1623–1641.
64. Sanner, M. F. (1999). Python: A programming language for software integration and development. *The Journal of Molecular Graphics and Modelling* 17, pp. 57–61.
65. Morris, G. M., Goodsell, D. S., Halliday, R. S., Huey, R., Hart, W. E., Belew, R. K. and Olson, A. J. (1998). Automated docking using a Lamarckian genetic algorithm and an empirical binding free energy function. *Journal of Computational Chemistry* 19(14), pp. 1639–1662.
66. Morris, G. M., Goodsell, D. S., Huey, R. and Olson, A. J. (1996). Distributed automated docking of flexible ligands to proteins: parallel applications of AutoDock 2.4. *Journal of Computer-aided Molecular Design* 10(4), pp. 293–304.

67. Shanno, D. F. (1970). Conditioning of quasi-Newton methods for function minimization. *Mathematics of Computation* 24(111), pp. 647–656.

68. Wessel, M. D., Jurs, P. C., Tolan, J. W. and Muskal, S. M. (1998). Prediction of human intestinal absorption of drug compounds from molecular structure. *Journal of Chemical Information and Computer Sciences* 38(4), pp. 726–735.

69. Clark, D. E. R. (1999). Calculation of polar molecular surface area and its application to the prediction of transport phenomena. 2. Prediction of blood–brain barrier penetration. *Journal of Pharmaceutical Sciences* 88, pp. 815–821.

70. Lipinski, C. A., Lombardo, F., Dominy, B. W. and Feeney, P. J. (2012). Experimental and computational approaches to estimate solubility and permeability in drug discovery and development settings. *Advanced Drug Delivery Reviews* 64, pp. 4–17.

71. Case, D., Darden, T., Cheatham, T., Simmerling, C., Wang, J., Duke, R., Luo, R., Crowley, M., Walker, R. C., Zhang, W., Merz, K., Wang, B., Hayik, S., Roitberg, A., Seabra, G., Kolossvry, I., Wong, K. F., Paesani, F., Vanicek, J., Wu, X., Brozell, S., Steinbrecher, T., Gohlke, H., Yang, L., Tan, C., Mongan, J., Hornak, V., Cui, G., Mathews, D., Seetin, M., Sagui, C., Babin, V. and Kollman, P. (2008). AMBER 10, University of California, San Francisco, CA.

72. Jorgensen, W. L., Chandrasekhar, J., Madura, J. D., Impey, R. W. and Klein, M. L. (1983). Comparison of simple potential functions for simulating liquid water. *The Journal of Chemical Physics* 79(2), pp. 926–935.

73. Darden, T., York, D. and Pedersen, L. (1993). Particle mesh Ewald: An Nlog(N) method for Ewald sums in large systems. *The Journal of Chemical Physics* 98, pp. 10089–10092.

74. Hockney, R. W., Goel, S. P. and Eastwood, J. (1974). Quit high resolution computer models of plasma. *Journal of Computational Physics* 14, pp. 148–158.

75. Uberuaga, B. P., Anghel, M. and Voter, A. F. (2004). Synchronization of trajectories in canonical molecular-dynamics simulations: Observation, explanation, and exploitation. *The Journal of Chemical Physics* 120, pp. 6363–6374.

76. Sindhikara, D. J., Kim, S., Voter, A. F. and Roitberg, A. E. (2009). Bad seeds sprout perilous dynamics: Stochastic thermostat induced trajectory synchronization in biomolecules. *Journal of Chemical Theory and Computation* 5, pp. 1624–1631.

77. Hess, B., Bekker, H., Berendsen, H. J. C. and Fraaije, J. G. E. M. (1997). LINCS: A linear constraint solver for molecular simulations. *Journal of Computational Chemistry* 18, pp. 1463–1472.

78. Berendsen, H. J. C., Postma, J. P. M., Vangunsteren, W. F., Dinola, A. and Haak, J. R. (1984). Molecular-dynamics with coupling to an external bath. *The Journal of Chemical Physics* 81, pp. 3684–3690.

79. Shrake, A. and Rupley, J. A. (1973). Environment and exposure to solvent of protein atoms. Lysozyme and insulin. *Journal of Molecular Biology* 79(2), pp. 351–371.

80. Sitkoff, D., Sharp, K. A. and Honig, B. (1994). Accurate calculation of hydration free energies using macroscopic solvent models. *The Journal of Physical Chemistry* 98(7), pp. 1978–1988.

81. Peng, S., Zhang, X., Lu, Y., Liao, X.-K., Lu, K., Yang, C., Liu, J., Zhu, W., Wei, D.-Q. (2016). mAMBER: A CPU/MIC Collaborated Parallel Framework for AMBER on Tianhe-2 Supercomputer. In: *Proceeding of IEEE International Conference on*

*Bioinformatics and Biomedicine*, BIBM, Shenzhen, China, December 15–18, 2016, pp. 651–657.

82. Talloen, W., Hochreiter, S., Bijnens, L., Kasim, A., Shkedy, Z., Amaratunga, D. and Göhlmann, H. (2010). Filtering data from high-throughput experiments based on measurement reliability. *Proceedings of the National Academy of Sciences of the United States of America* 107(46), p. E173.

83. Talloen, W., Clevert, D. A., Hochreiter, S., Amaratunga, D., Bijnens, L., Kass, S. and Göhlmann, H. W. (2007). I/NI-calls for the exclusion of non-informative genes: a highly effective filtering tool for microarray data. *Bioinformatics* 23(21), pp. 2897–2902.

84. Wouters, L., Göhlmann, H. W., Bijnens, L., Kass, S. U., Molenberghs, G. and Lewi, P. J. (2003). Graphical exploration of gene expression data: A comparative study of three multivariate methods. *Biometrics* 59(4), pp. 1131–1139.

85. Smyth, G. K., Ritchie, M., Thorne, N. and Wettenhall, J. (2005). LIMMA: linear models for microarray data. In Bioinformatics and Computational Biology Solutions Using R and Bioconductor. Statistics for Biology and Health.

86. Iskar, M., Zeller, G., Blattmann, P., Campillos, M., Kuhn, M., Kaminska, K. H., Runz, H., Gavin, A. C., Pepperkok, R., Van Noort, V. and Bork, P. (2013). Characterization of drug-induced transcriptional modules: towards drug repositioning and functional understanding. *Molecular Systems Biology* 9(1), p. 662.

87. Hochreiter, S., Bodenhofer, U., Heusel, M., Mayr, A., Mitterecker, A., Kasim, A., Khamiakova, T., Van Sanden, S., Lin, D., Talloen, W. and Bijnens, L. (2010). FABIA: factor analysis for bicluster acquisition. *Bioinformatics* 26(12), pp. 1520 1527.

88. Hochreiter, S. and Obermayer, K. (2006). Support vector machines for dyadic data. *Neural Computation* 18(6), pp. 1472–1510.

Chapter 6

# A Hybrid Approach Integrating Model-Based Method and Gene Functional Similarity for Cluster Analysis of RNA-Seq Data

Ming-Han Chan*, Pin-Chen Chou*, Rong-Ming Chen*, and Rouh-Mei Hu[†]

*National University of Tainan, Taiwan
[†]Asia University, Taiwan

## 6.1. Introduction

Transcriptome is the set of RNA molecules in a cell. Study of transcriptome can help us to get insight into the biological activities of the cell and understand the development of diseases. Microarray is the first powerful high-throughput approach to investigate the transcription levels of interesting genes in a genome. However, this method has some limitations, especially in the high-noise and low-resolution. By using the high-throughput sequencing capacity of next-generation sequencing (NGS) method, RNA sequencing (RNA-seq) directly reveals the sequence and abundance of transcripts and has substituted the traditional hybridization-based microarray approach in transcriptome studies. Compared to microarray, the high-resolution character of RNA-seq data provides information of not only the gene transcription level, but also the alternative splicing, allelic specific transcription, and new genes. These advantages made RNA-seq a leading technology over the previous methods in transcriptomic studies. Due to the large scale and high complexity of RNA-seq data, new methods have to be developed for data analysis [1–10].

Cluster analysis has been widely used in microarray data study. Assuming that genes with similar expression patterns may have identical or similar functions, or be collectively involved in some transcriptional regulatory networks. An approach called as the guilt-by-association principle [11, 12]. Based on the gene expression level, we may group co-expressional genes from the microarray data and assign new function to poorly annotated genes or identify new genes that might be involved in specific diseases.

Heuristic k-means clustering algorithm and probabilistic model-based method have been applied in RNA-seq data analysis. Li *et al.* [13] used the heuristic k-means clustering algorithm to investigate the transcriptional regulatory network associated with the development of C4 photosynthesis in maize leaf. Although this algorithm is simple and can be easily implemented, a study on microarray data has showed that heuristic methods are not as effective as probabilistic model-based methods [14]. Due to the discrete counts and skewed properties of RNA-seq data, a Gaussian mixture model designed for microarray data analysis is unsuitable for data analysis. Instead, Poisson or negative binomial (NB) distributions in probabilistic models have been modified for RNA-seq study [15–17]. Witten [18] proposed a hierarchical clustering algorithm by using a Poisson-based probabilistic model and a new dissimilarity measure to cluster RNA sequence data. Their evaluation results demonstrated that Poisson model is appropriate for RNA sequence classification and clustering. Si *et al.* [8] applied a probabilistic model-based clustering algorithm that fit RNA-seq data to Poisson and NB distributions. This method has been implemented to analyze the maize leaf dataset published by Li *et al.* [13] and showed a better clustering result than the original k-means clustering algorithm and hierarchical clustering methods.

Methods have been developed for time series RNA-seq data analysis [1, 19]. Time-course analysis of RNA-seq data focuses on temporal dynamic changes in gene expression, and the dynamic correlation between time points is crucial. Oh *et al.* [19] discussed several methods that can be used to model time-course RNA-seq data, such as the statistical evolutionary trajectory index, autoregressive time-lagged regression, and the hidden Markov model. They used real datasets and simulation studies to demonstrate the utility of these dynamic methods within temporal analysis. Äijö *et al.* [1] used non-parametric Gaussian processes to interpret the temporal correlation in gene expression and used a NB distribution to construct a probabilistic model for RNA-seq read counting.

Although co-expression analysis is a popular method in gene function prediction. Genes with a similar expression probability distribution or pattern may not necessarily have identical or similar functions. Equally, genes with identical or similar functions may not have a similar probability distribution or pattern of expression. Therefore, both gene functional similarity and the probability distribution/patterns of gene expression are crucial factors in cluster analysis and incorporating both factors should improve the correlation of gene functions during a cluster analysis. This study integrated the probability distribution information of gene expression and gene functional similarities to conduct the cluster analysis of RNA-seq data.

The remainder of this chapter proceeds as follows: methodology and procedures, experimental design and analysis, evaluation and comparison of the clustering effectiveness, and conclusion.

## 6.2. Methodology

A flowchart describing the cluster analysis of RNA-seq data is presented in Fig. 6.1. The central idea was to calculate the model-based and semantic-based gene pair distances using gene clusters corresponding to probability distributions and gene functional similarity, respectively. A hybrid gene distance matrix was constructed with feature weights calculated using a bootstrapping method. Distance-based clustering was then conducted.

This section is divided into five subsections. Subsection 6.2.1 discusses how to convert the gene clusters obtained using model-based clustering into gene pair distances, which was named the *model-based gene pair distance*; Subsection 6.2.2 introduces how to compute the *semantic-based gene pair distance* using gene functional similarity; Subsection 6.2.3 discusses how to construct the hybrid gene distance matrix featuring both the model-based

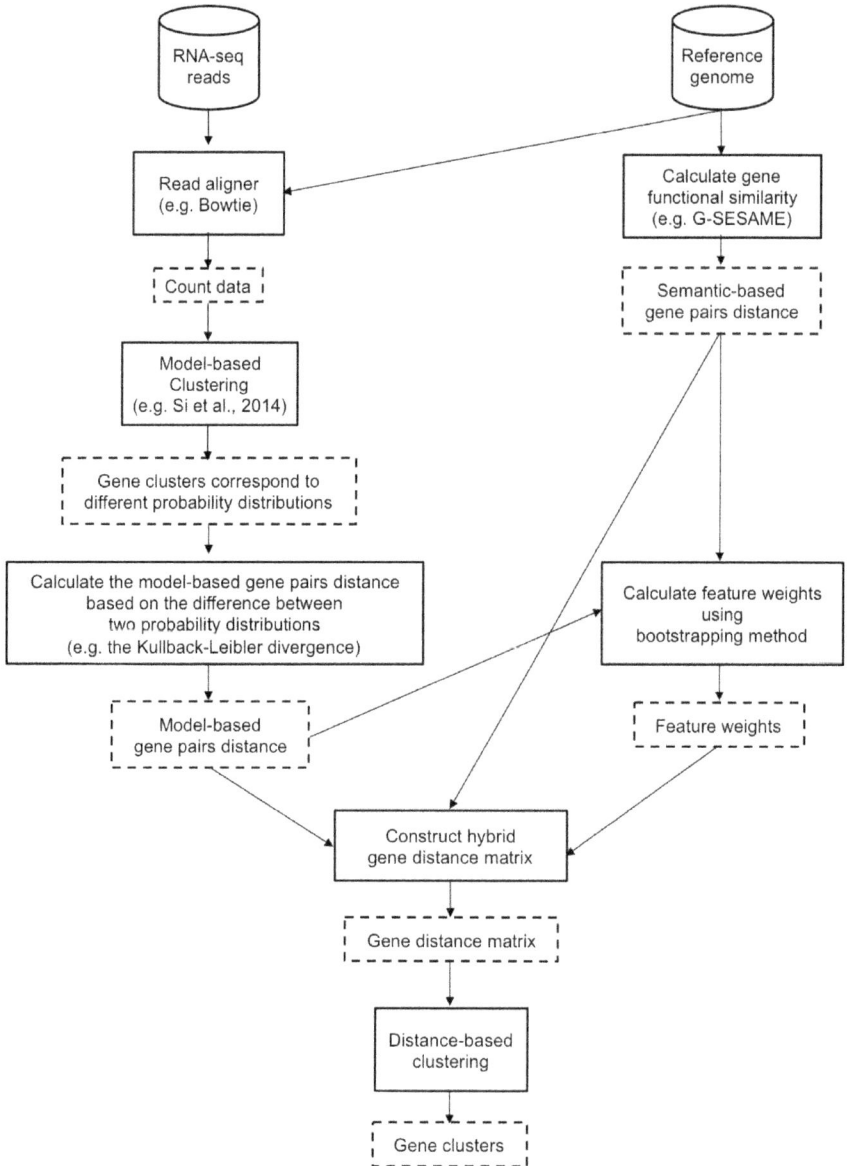

Fig. 6.1.	Flowchart of cluster analysis of RNA-seq data.

and semantic-based gene pair distances; Subsection 6.2.4 introduces how the feature weights of the model-based and semantic-based gene pair distances were selected; and Subsection 6.2.5 describes the gene pair distance-based cluster analysis.

## 6.2.1. *Computation of the Model-based Gene Pair Distances Using Gene Clusters Corresponding to Probability Distributions*

To obtain the model-based gene pair distances, we first had to perform model-based clustering of the RNA-seq data. This study employed the probabilistic model and clustering methodology proposed by Si *et al.* [8] and herein described briefly the methodology, and then discusses how the present study calculated the model-based gene pair distances using the clustering results through the probability distribution model.

### 6.2.1.1. *Clustering model-based gene expression data*

The gene expression data in this study consisted of a large number of short sequence reads generated by high-throughput RNA-seq. To obtain the count data of the reads, they were mapped to the reference genome using a Bowtie read aligner.

As Si *et al.* [8] described, probability models in previous studies of RNA-seq data consisted of Poisson and NB distributions. The Poisson distribution has been proven to be adequate for RNA-seq data when only technical replicates are involved [15, 16]. If the experiment is considered a biological replicate experiment, RNA-seq data may exhibit more variability than expected from a Poisson distribution, namely the overdispersion phenomenon. Previous studies have revealed that the NB distribution is a more appropriate probabilistic model than the Poisson distribution [55]. Therefore, Si *et al.* [8] used the parametrization proposed by Robinson and Smyth [20].

Gene expression data clustering using a probabilistic model assumes that the gene expression data in the mixture model are sampled from the parent population. Each probability distribution submodel corresponds to a gene cluster, and the number of clusters must be determined or configured through other clustering validity methods. Previous studies have employed clustering validity methods such as Akaike's information criterion [21], the Bayesian information criterion [22], the Hubert gamma statistic [23], the Dindex [24], and the hybrid-hierarchical [8, 25, 26]. The most commonly used hybrid probability distribution for model-based analysis of microarray gene expression data has been the multivariate Gaussian mixture model [14, 27].

Through the defined latent variable or hidden data of the cluster membership function, the optimal cluster membership function can be iteratively estimated using the well-known expectation–maximization (EM) algorithm [8, 27–29]. Because Si *et al.* [8] explained in detail the Poisson and NB mixture model-based cluster analysis of RNA-seq gene expression data using the EM algorithm, the present study only outlined the concepts of their methodology.

### 6.2.1.2.  *Computation of model-based gene pair distances based on the difference between two probability distributions*

Clustering of gene expression data was conducted using the aforementioned probabilistic model proposed in previous studies. Each gene was grouped into one of the $K$ clusters, and the genes in one cluster collectively belonged to one probability distribution submodel. The distance between genes was calculated using this probabilistic model, named the *model-based gene pair distance*. For genes belonging to the same probability distribution submodel, the gene pair distance was defined as zero; for genes belonging to different submodels (e.g. two genes belonging to two different probability distribution functions $f_1$ and $f_2$), the gene pair distance was defined as the distance between $f_1$ and $f_2$.

One of the most commonly used methods of measuring the distance between different probability distributions has been the Kullback–Leibler (KL) divergence from the field of information theory [30–34]. The present study used the KL divergence to compute the difference between two probability distributions as the distance between the two genes with different probability distributions, called the *KL-based gene pair distance* or *model-based gene pair distance*. Assume $f_1$ and $f_2$ are two different probability distribution functions in a discrete probability distribution. Because the KL divergence is an asymmetric equation, the KL divergence from $f_1$ to $f_2$ is defined as

$$D_{\text{KL}}(f_1 \| f_2) = \sum_i f_1(i) \ln \left( \frac{f_1(i)}{f_2(i)} \right) \tag{6.1}$$

where if $f_2(i) = 0$, $f_1(i) = 0$; similarly, the KL divergence from $f_2$ to $f_1$ is defined as

$$D_{\text{KL}}(f_2 \| f_1) = \sum_i f_2(i) \ln \left( \frac{f_2(i)}{f_1(i)} \right) \tag{6.2}$$

where if $f_1(i) = 0$, $f_2(i) = 0$; and the symmetric KL divergence is defined as

$$D_{\text{KL}}(f_1, f_2) = D_{\text{KL}}(f_1 \| f_2) + D_{\text{KL}}(f_2 \| f_1). \tag{6.3}$$

On the basis of this KL divergence, the model-based gene pair distance was defined as follows. Suppose gene $g_1$ and $g_2$ belong to $f_1$ and $f_2$, respectively. Then the model-based distance between $g_1$ and $g_2$ is

$$D_{\text{MB}}(g_1, g_2) = D_{\text{KL}}(f_1, f_2). \tag{6.4}$$

### 6.2.2. *Computation of Gene Functional Similarity*

This section describes how to compute the distance between gene pairs using gene functional similarity. This study employed gene ontology (GO) semantic similarity to measure the *semantic-based gene pair distance* or *information content-based gene pair distance*. GO is an annotation database of genetic functions published by the Gene Ontology Consortium [35], and is hierarchically structured as a directed acyclic graph (DAG). Each node has a word or string of words as the GO term for annotation.

Numerous studies on GO semantic similarity have been published. The two most common methods consist of topology-based and annotation-based similarity. The topology-based method calculates gene semantic similarities using GO DAG [36, 37], whereas the annotation-based method measures similarities statistically using GO terms [38–40]. The present study used one well-known topology-based [40] and two annotation-based methods [41] to compute semantic-based gene pair distances in which $D_{SB}(g_1, g_2)$ denotes the distance between genes $g_1$ and $g_2$.

### 6.2.3. Construction of Model- and Semantic-based Hybrid Gene Distance Matrix

A gene pair distance matrix is generally defined before cluster analysis is performed. This section describes how to obtain a hybrid gene pair distance matrix combining model- and semantic-based distances using feature weights. Let $D(g_1, g_2)$ denote the integrated distance between genes $g_1$ and $g_2$. Suppose $r$ and $1 - r$ are the weights of the model- and semantic-based distances, respectively. Then the integrated distance is given by

$$D(g_1, g_2) = r D_{MB}(g_1, g_2) + (1 - r) D_{SB}(g_1, g_2). \qquad (6.5)$$

Hence, the integrated gene distance matrix

$$[D(g_1, g_2)]_{G \times G} = \begin{bmatrix} D(1, 1) & \cdots & D(1, G) \\ \vdots & \ddots & \vdots \\ D(G, 1) & \cdots & D(G, G) \end{bmatrix} \qquad (6.6)$$

where $g_1, g_2 = 1, \ldots, G$, and $G$ is the number of genes to be analyzed.

### 6.2.4. Determination of the Feature Weights of Model- and Semantic-based Gene Distances

Next, the feature weight $r$ had to be determined. This study used the effective bootstrapping method, commonly used in previous studies, to compute the feature weight of the gene distance [42–44]. The concept and steps of the bootstraping method are described as follows, but the detailed equations are not repeated here. Similar mathematical equations can be referenced in the studies published by Hung *et al.* [45] and Chen and Chen [46].

Because the variance of model- and semantic-based gene distances had to be estimated and normalized before the bootstrapping method could be employed, the present study used Pearson's coefficient to compute the coefficient of variation and the feature weight. However, because of the stochastic nature of gene expression data, the sample mean of the weight is generally used in practice. To estimate the sample mean weight, a certain amount of sample data is required. This study employed the commonly used balanced bootstrapping method, in which replicates of sampling data that were comparable to the parent population in numbers were generated using random sampling with replacement [47].

### 6.2.5. *Gene Distance-based Cluster Analysis*

After the model- and semantic-based hybrid gene distance matrix was constructed, cluster analysis was conducted using the distanced-based algorithm. Famous clustering methods include k-means, hierarchical, self-organizing map (SOM), fuzzy c-means [48–51], and other effective clustering methods. The present study employed the k-medoids algorithm as the clustering method because of its robustness to noise and outliers.

### 6.3. Research Design and Analysis of Results

### 6.3.1. *Research Design*

To assess the clustering effectiveness of the model- and semantic-based hybrid methods, this study used the same model-based gene RNA-seq count data as the data employed by Si *et al.* [8] for experimental verification.

The goal of clustering is to group genes with identical or similar functions. Thus, Si *et al.* [8] used the maize leaf transcriptome proposed by Li *et al.* [13]. First, functional categories that contained <5 or >500 genes were excluded, and 306 non-overlapping categories containing 5,002 genes were selected. Gene annotations were obtained from MapMan [52] as described by Li *et al.* [13]. Among the 5,002 selected genes, only 2,220 of them exhibited count data and these were distributed among 293 non-overlapping categories. Therefore, the present study conducted experimental verification using these 2,220 genes with RNA-seq count data.

A comprehensive analysis was conducted to verify the clustering effectiveness of the model- and semantic-based hybrid methods. In experimental design, the feature weight was selected by using the bootstrapping method. The number of clusters ranged from $k = 20$ to $k = 100$, the number increased by 10 in each successive experiment. In the model-based method, the NB and Poisson distributions were employed as the distribution function, and the expectation–maximization (EM) algorithm was employed to estimate the optimal parameters [8]. The gene semantic similarities were calculated using Resnik, simGIC, and simUI [40, 41]. The main structure of the GO consisted of biological processes (BP), cellular components (CC), and molecular functions (MF) [35]. The results from the hybrid method proposed in the present study were compared with those of both model-based and semantic-based methods, along with those that used the conventional k-means and SOM clustering methods.

### 6.3.2.  *Evaluation Methodology*

To ensure consistency, this study employed the normalized mutual information (NMI) used by Si *et al.* [8] as the quantitative measure. Mutual information (MI) is used in information theory to measure the amount of information one random variable contains about another. The application of MI is equivalent to the concordance between the result of the quantization experiment (Partition $A$) and the true partition result (Partition $B$). A high MI value suggests strong dependence between these two partitions, with low MI indicating weak dependence. Because there is no upper bound to the MI value and the lowest possible value is zero, the NMI value, ranging from 0 to 1, is often used to simplify comparisons. Suppose $I(A, B)$ denotes the MI between Partitions $A$ and $B$, and $H(A)$ and $H(B)$ represent the entropies of Partitions $A$ and $B$, respectively. Then, the NMI between Partitions $A$ and $B$ is defined as [53]:

$$\text{NMI}(A, B) = \frac{I(A, B)}{\sqrt{H(A)H(B)}}. \tag{6.7}$$

Some other normalization methods involve arithmetic and geometric means. Because $H(A) = I(A, A)$, Strehl and Ghosh [53] preferred the geometric mean because of its analogy with a normalized inner product in Hilbert space. The limited clustering data of Partitions $A$ and $B$ were required for computing NMI$(A, B)$. Suppose $K(A)$ and $K(B)$ denote the number of clusters in Partitions $A$ and $B$, respectively; $N_i(A)$ and $N_i(B)$ are the number of genes in the cluster $C_i$ in Partition $A$ and cluster $C_j$ in Partition $B$, respectively; and $N_{ij}(A, B)$ denote the number of genes appearing in both the cluster $C_i$ in Partition $A$ and cluster $C_j$ in Partition $B$. The estimated NMI value can then be given as follows:

$$\hat{\text{NMI}}(A, B) = \frac{\sum_{i=1}^{K(A)} \sum_{j=1}^{K(B)} N_{ij}(A, B) \log \left( \frac{G \cdot N_{ij}(A,B)}{N_i(A)N_j(B)} \right)}{\sqrt{\left( \sum_{i=1}^{K(A)} N_i(A) \log \left( \frac{N_i(A)}{G} \right) \right) \left( \sum_{j=1}^{K(B)} N_j(B) \log \left( \frac{N_j(B)}{G} \right) \right)}}.$$

(6.8)

### 6.3.3. *Experimental Results and Analysis*

To verify the suitability of the experimental results, this section presents and examines the effect of mixing the two feature weights on the clustering effectiveness of the hybrid model- and semantic-based method. The remainder of this section is divided into two parts. First, the feature weights were selected using the bootstrapping method and were used to demonstrate the clustering results of the hybrid mixture model. Second, the optimal number of clusters (i.e. the cluster validity) was then analyzed.

6.3.3.1. *Clustering results with the bootstrapped feature weights*

To facilitate comparative analysis, the experimental result performance indicator, NMI, was plotted under various settings for the number of clusters. First, NB was used as the distribution function and EM as the algorithm for estimating the optimal parameters. Figures 6.2–6.4 illustrate the clustering effectiveness calculated by Resnik, simGIC, and simUI with BP, CC, and MF as the GO, respectively. The feature weights of the mixture model were selected using the bootstrapping method. The mixture model yielded more favorable results than either the model-based or semantic-based method alone when the number of clusters exceeded around 50. In addition, the mixture model performed more favorably than either the k-means or SOM method when the number of clusters exceeded around 70. The optimal number of clusters is discussed in Section 6.3.3.2. Second, the experimental results with the Poisson distribution are examined, Figs. 6.5–6.7 clearly

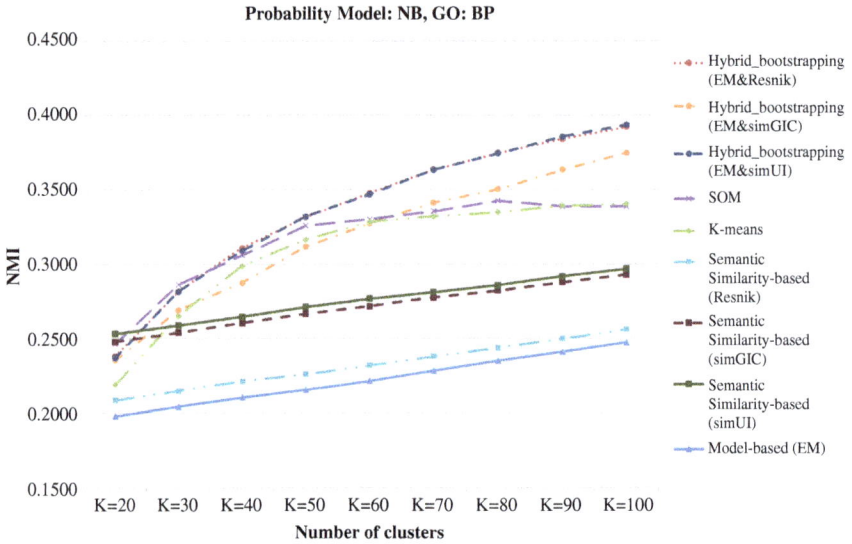

Fig. 6.2. Comparison of performance indicator NMI for the probability model NB with BP as the GO.

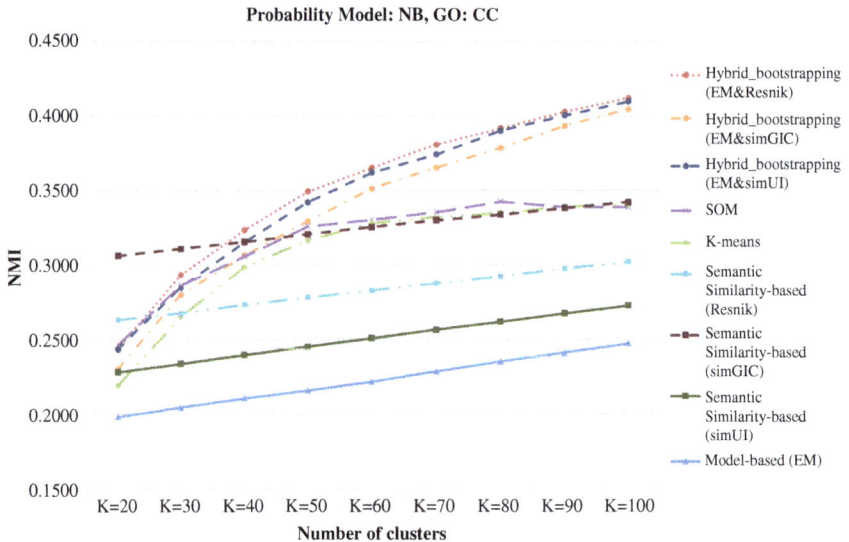

Fig. 6.3. Comparison of performance indicator NMI for the probability model NB with CC as the GO.

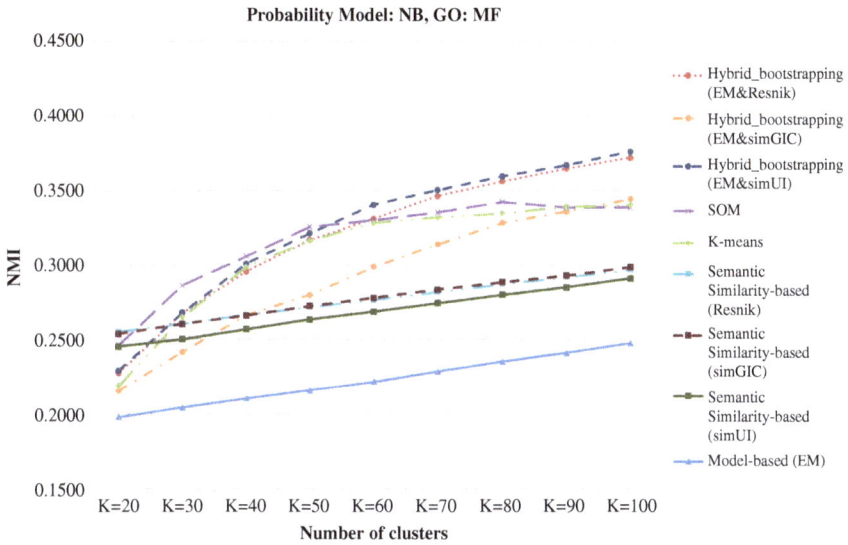

Fig. 6.4. Comparison of performance indicator NMI for the probability model NB with MF as the GO.

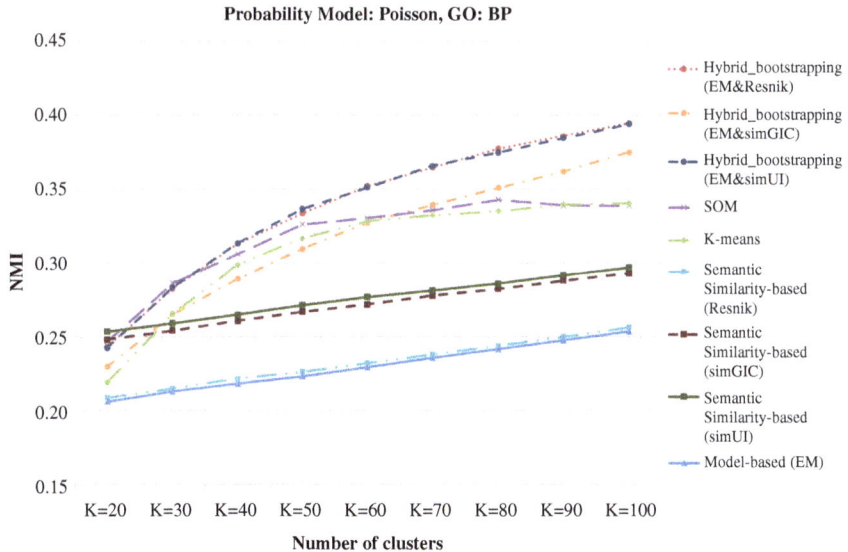

Fig. 6.5. Comparison of performance indicator NMI for the probability model Poisson with BP as the GO.

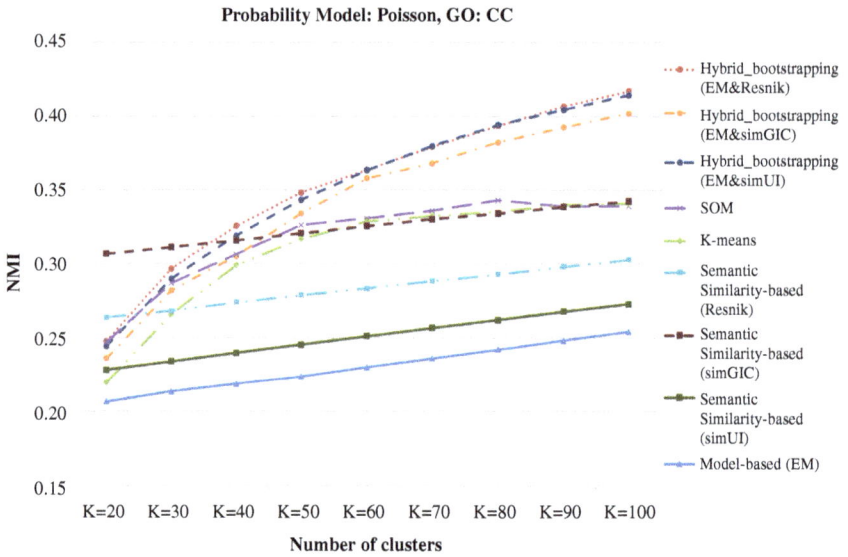

Fig. 6.6. Comparison of performance indicator NMI for the probability model Poisson with CC as the GO.

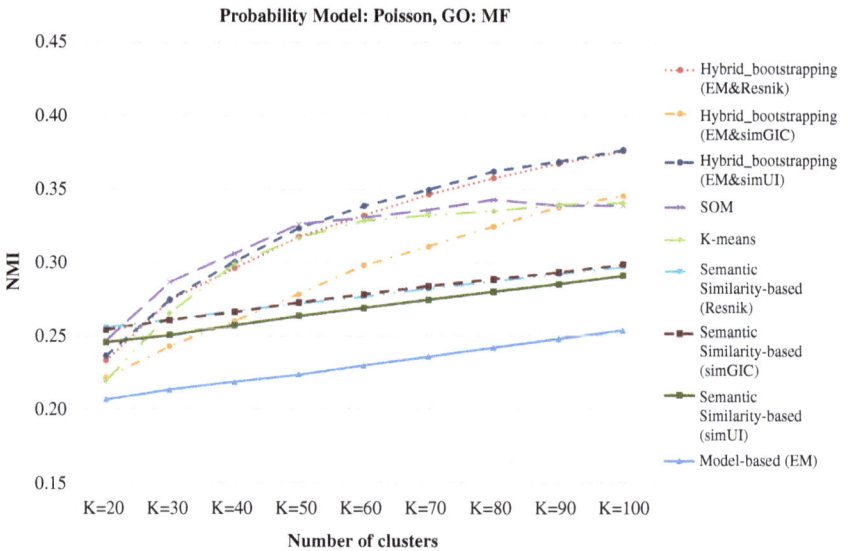

Fig. 6.7. Comparison of performance indicator NMI for the probability model Poisson with MF as the GO.

demonstrate that the mixture model also yielded more favorable results than either the model-based or semantic-based method alone when the number of clusters exceeded around 50. In addition, the mixture model also performed more favorably than either the k-means or SOM method when the number of clusters exceeded around 70.

### 6.3.3.2. *Cluster validity analysis*

This study employed two common indicators — the Dindex and Hubert gamma statistics — for cluster validity analysis using the R Package NbClust [54]. Figures 6.8 and 6.9 illustrate the Dindex obtained through the average hierarchical agglomerative clustering (HAC) and median HAC methods, respectively, whereas Figs. 6.10 and 6.11 present the corresponding Hubert gamma statistics obtained through the same methods. The optimal number of clusters were selected using the positions of the second difference plots with significant peaks (Figs. 6.8–6.11). These results revealed

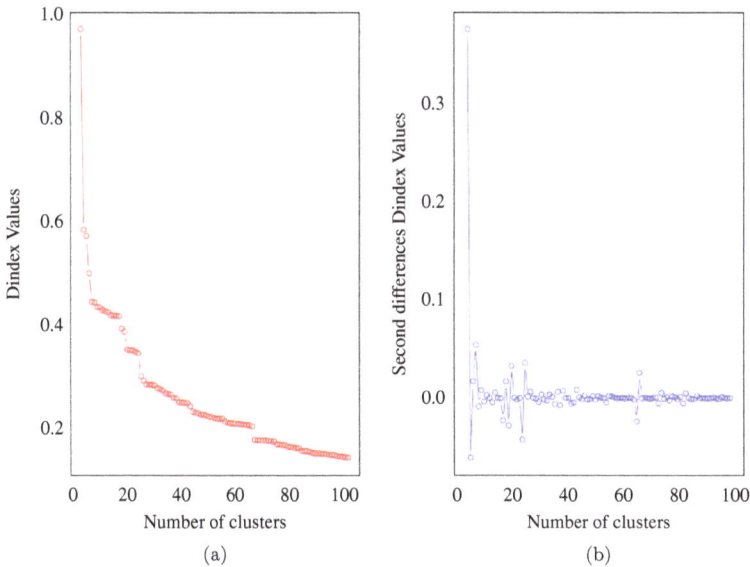

Fig. 6.8. (a) Dindex obtained using the average HAC method; (b) second difference plot for Dindex.

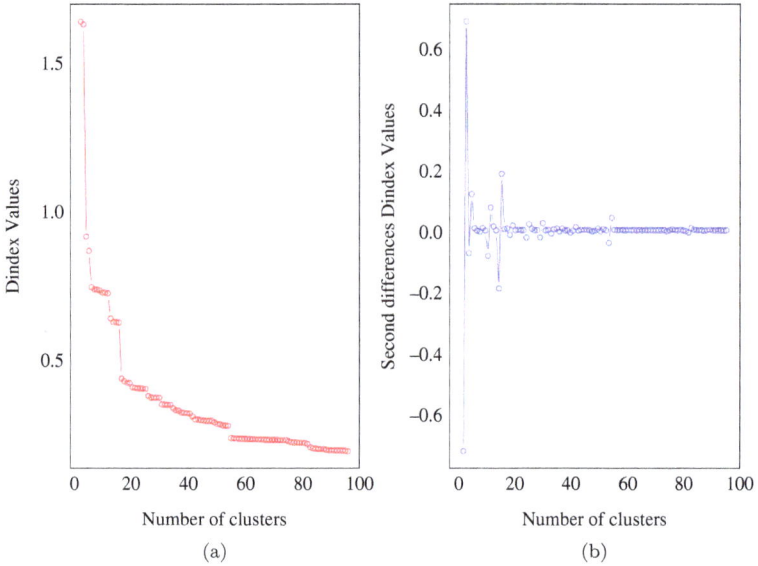

Fig. 6.9.   (a) Dindex obtained using the median HAC method; (b) second difference plot for Dindex.

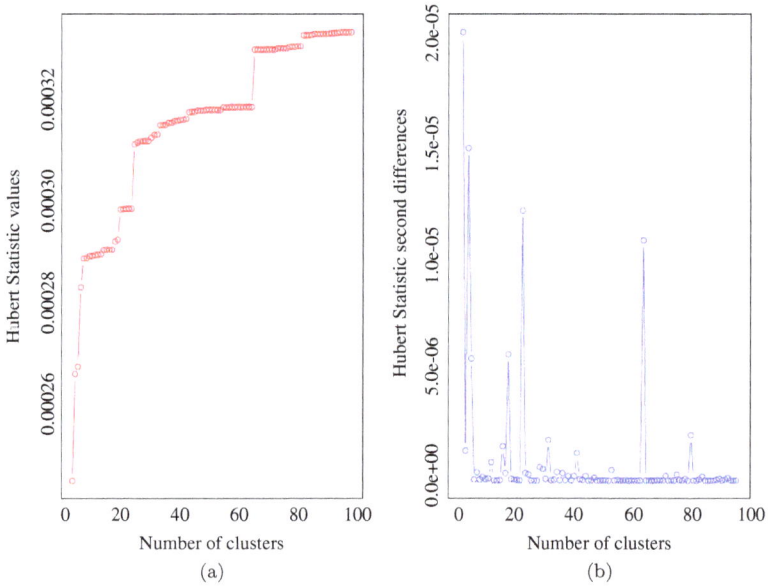

Fig. 6.10.   (a) Hubert Gamma Statistic obtained using the average HAC method; (b) second difference plot for Hubert Gamma Statistic.

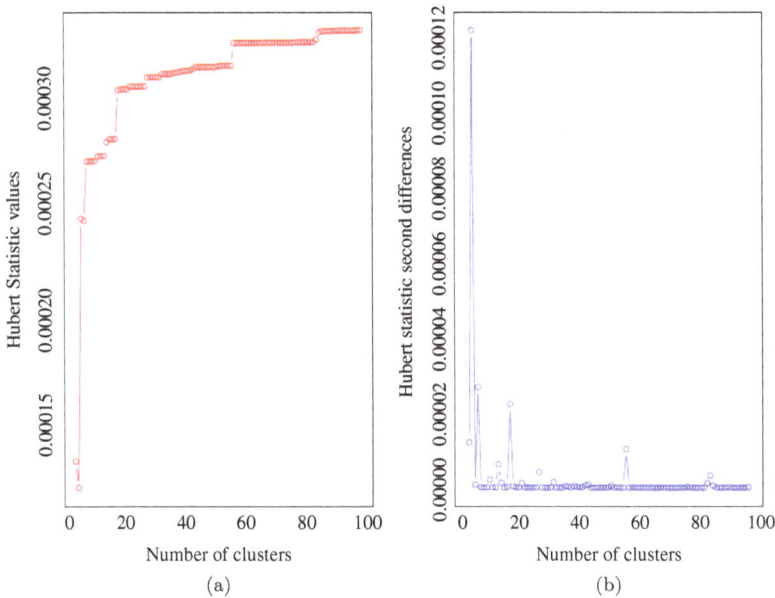

Fig. 6.11. (a) Hubert Gamma Statistic obtained using the median HAC method; (b) second difference plot for Hubert Gamma Statistic.

that the number of clusters in Section 6.3.3.1 should exceed 80 for the reference to be reliable when the experimental results were compared with those using the conventional $k$-means and SOM clustering methods.

## 6.4. Conclusion and Subsequent Research

This study proposed a hybrid of the model-based method and gene functional similarity for cluster analysis of RNA-seq data. The central idea was to compute the model- and semantic-based gene distances using gene clusters of probability distributions generated by gene expression data and gene semantic similarity, respectively. A hybrid gene distance matrix was constructed using the bootstrapping method, and clustering was conducted using distance-based algorithms. Finally, real RNA-seq data were used for experimental verification and the commonly used NMI was employed as the clustering performance indicator. The mixture model proposed in this study yielded significantly more favorable clustering results than either the model-based or semantic-based clustering method alone when the number of clusters exceeded a reasonable value. In addition, the mixture model was compared with two commonly used clustering algorithms (k-means and SOM) and discovered to yield more favorable clustering results when the number of clusters exceeded a reasonable value. This study can serve as a reference for subsequent research relating to the precise reconstruction of genetic regulatory networks.

## Acknowledgments

Special thanks to the Ministry of Science and Technology of Taiwan for the research grants that funded this study (Grant Nos: MOST 104-2221-E-024-018 and MOST 105-2221-E-024-020).

## References

1. Äijö, T., Butty, V., Chen, Z., Salo, V. and Tripathi, S. (2014). Methods for time series analysis of RNA-seq data with application to human Th17 cell differentiation. *Bioinformatics* 30, ISMB, pp. i113–i120.
2. Cao, Y., Zhu, J., Jia, P. and Zhao, Z. (2017). scRNASeqDB: A database for RNA-seq based gene expression profiles in human single cells. *Genes* 8, p. 368.
3. Cheng, X. and Jin, V. X. (2018). An introduction to integrative genomics and systems medicine in cancer. *Genes* 9, p. 37.
4. Conesa, A. *et al.* (2016). A survey of best practices for RNA-seq data analysis. *Genome Biology* 17, pp. 1–19.
5. Li, B., Ruotti, V., Stewart, R. M., Thomson, J. A. and Dewey, C. N. (2010). RNA-seq gene expression estimation with read mapping uncertainty. *Bioinformatics* 26, pp. 493–500.
6. Marguerat, S., Wilhelm, B. T. and Bähler, J. (2008). Next-generation sequencing: Applications beyond genomes. *Biochemical Society Transactions* 36, pp. 1091–1096.
7. Metzker, M. (2010). Sequencing technologies — the next generation. *Nature Reviews Genetics* 11, pp. 31–46.
8. Si, Y., Liu, P. Li, P. and Brutnell, T. P. (2014). Model-based clustering for RNA-seq data. *Bioinformatics* 30, pp. 197–205.
9. Wang, Z., Gerstein, M. and Snyder, M. (2009). RNA-seq: A revolutionary tool for transcriptomics. *Nature Reviews Genetics* 10, pp. 53–67.
10. Wang, L., Li, P. and Brutnell, T. P. (2010). Exploring plant transcriptomes using ultra high-throughput sequencing. *Briefings in Functional Genomics* 9, pp. 118–128.
11. Oliver, S. (2000). Proteomics: Guilt-by-association goes global. *Nature* 403, pp. 601–603.

12. Wolfe, C. J., Kohane, I. S. and Butte1, A. J. (2005). Systematic survey reveals general applicability of "guilt-by-association" within gene coexpression networks. *BMC Bioinformatics* 6, pp. 227–236.
13. Li, P., Ponnala, L., Gandotra, N., Wang, L., Si, Y., Tausta, S., Kebrom, T., Provart, N., Patel, R., Myers, C., Reidel, E., Turgeon, R., Liu, P., Sun, Q., Nelson, T. and Brutnell, T. (2010). The developmental dynamics of the maize leaf transcriptome. *Nature Genetics* 42, pp. 1060–1067.
14. Yeung, K., Fraley, C., Murua, A., Faftery, A. and Ruzzo, W. (2001). Model-based clustering and data transformations for gene expression data. *Bioinformatics* 17, pp. 977–987.
15. Bullard, J., Purdom, E., Hansen, K. and Dudoit, S. (2010). Evaluation of statistical methods for normalization and differential expression in mRNA-seq experiments. *BMC Bioinformatics* 11, p. 94.
16. Marioni, J. C. *et al.* (2008). RNA-seq: An assessment of technical reproducibility and comparison with gene expression arrays. *Genome Research* 18, pp. 1509–1517.
17. Robinson, M. D., McCarthy, D. J. and Smyth, G. K. (2010). Edger: a bioconductor package for differential expression analysis of digital gene expression data. *Bioinformatics* 26, pp. 139–140.
18. Witten, D. (2011). Classification and clustering of sequencing data using a Poisson model. *Annals of Applied Statistics* 5, pp. 2493–2518.
19. Oh, S., Song, S., Grabowski, G., Zhao, H. and Noonan, J. P. (2013). Time series expression analyses using RNA-seq: A statistical approach. *BioMed Research International* 2013, Article ID 203681, 16 pages.
20. Robinson, M. D. and Smyth, G. K. (2008). Small-sample estimation of negative binomial dispersion, with applications to sage data. *Biostatistics* 9, pp. 321–332.
21. Akaike, H. (1974). A new look at the statistical model identification. *IEEE Trans. Automat. Contr.* 19, pp. 716–723.
22. Schwarz, G. (1978). Estimating the dimensions of a model. *Annals of Statistics* 6, pp. 461–464.
23. Hubert, L. J. and Arabie, P. (1985). Comparing partitions. *Journal of Classification* 2, pp. 193–218.
24. Lebart, L., Morineau, A. and Piron, M. (2000). *Statistique Exploratoire Multidimensionnelle* (Dunod, Paris).
25. Vaithyanathan, S. and Dom, B. (2000). Model-based hierarchical clustering. *Proceedings of the 16th Conference on Uncertainty in Artificial Intelligence*, pp. 599–608.
26. Zhong, S. and Ghosh, J. (2003). A unified framework for model-based clustering. *Journal of Machine Learning Research* 4, pp. 1001–1037.
27. Fraley, C. and Raftery, A. (2002). Model-based clustering, discriminant analysis, and density estimation. *Journal of the American Statistical Association* 97, pp. 611–631.
28. Dempster, A. P., Laird, N. M. and Rubin, D. B. (1977). Maximum likelihood from incomplete data via the EM algorithm. *Journal of the Royal Statistical Society, Series B* 39, pp. 1–38.
29. McLachlan, G. (1997). On the em algorithm for overdispersed count data. *Statistical Methods in Medical Research* 6, pp. 76–98.

30. Kuijf, H. J., van Veluw, S. J., Geerlings, M. I., Viergever, M. A., Biessels, G. J. and Vincken, K. L. (2014). Automatic extraction of the midsagittal surface from brain MR images using the Kullback–Leibler measure. *Neuroinformatics* 12, pp. 395–403.
31. Kullback, S. (1959). *Information Theory and Statistics* (Wiley, New York).
32. Kullback, S. and Leibler, R. A. (1951). On information and sufficiency. *Annals of Mathematical Statistics* 22, pp. 79–86.
33. Puspitasari, F., Volkau, I., Ambrosius, W. and Nowinski, W. L. (2009). Robust calculation of the midsagittal plane in CT scans using the Kullback–Leibler's measure. *International Journal of Computer Assisted Radiology and Surgery* 4, pp. 535–547.
34. Zheng, Q., Lu, Z., Yang, W., Zhang, M., Feng, Q. and Chen, W. (2013). A robust medical image segmentation method using KL distance and local neighborhood information. *Computers in Biology and Medicine* 43, pp. 459–470.
35. Ashburner, M., Ball, C. A., Blake, J. A., Botstein, D., Butler, H., Cherry, J. M. and Eppig, J. T. (2000). Gene ontology: Tool for the unification of biology. *Nature Genetics* 25, pp. 25–29.
36. Mazandu, G. K. and Mulder, N. J. (2012). A topology-based metric for measuring term similarity in the gene ontology. *Advances in Bioinformatics* Article ID 975783, 17 pages.
37. Zhang, P., Jinghui, Z., Huitao, S., Russo, J., Osborne, B. and Buetow, K. (2006). Gene functional similarity search tool (GFSST). *BMC Bioinformatics* 7, p. 135.
38. Jiang, J. J. (1997). Conrath DW: Semantic similarity based on corpus statistics and lexical taxonomy. *Proc. 10th International Conference on Research in Computational Linguistics*, pp. 19–33.
39. Lin, D. (1998). An information-theoretic definition of similarity. *Proc. 15th International Conference on Machine Learning*, pp. 296–304.
40. Resnik, P. (1999). Semantic similarity in a taxonomy: An information-based measure and its application to problems of ambiguity in natural language. *Journal of Artificial Intelligence Research* 11, pp. 95–130.
41. Wang, J. Z., Du, Z., Payattakool, R., Yu, P. S. and Chen, C. F. (2007). A new method to measure the semantic similarity of GO terms. *Bioinformatics* 23, pp. 1274–1281.
42. Efron, B. (1979). Bootstrap methods: Another look at the jackknife. *Annals of Statistics* 7, pp. 1–26.
43. Efron, B. and Tibshirani, R. (1986). Bootstrap methods for standard errors, confidence intervals, and other measures of statistical accuracy. *Statistical Science* 1, pp. 54–77.
44. Efron, B. and Tibshirani, R. (1993). *An Introduction to Bootstrap* (Chapman and Hall, New York).
45. Hung, W. L., Yang, M. S. and Chen, D. H. (2008). Bootstrapping approach to feature-weight selection in fuzzy c-means algorithms with an application in color image segmentation. *Pattern Recognition Letters* 29, pp. 1317–1325.
46. Chen, L. C. and Chen, R. M. *et al.* (2014). A bootstrapping approach to distance-based clustering analysis of gene expression data with incorporation of GO term based gene similarity. Proc. 19th Conference on Technologies and Applications of Artificial Intelligence, TAAI (in Chinese).
47. Chou, H. Y. (2004). *Investigation of Bootstrap and Its Applications* (Master Thesis, National Central University, Taiwan) (in Chinese).

48. Carr, D. B., Somogyi, R. and Michaels, G. (1997). Templates for looking at gene expression clustering. *Statistical Computing & Statistical Graphics Newsletter* 8, pp. 20–29.
49. Jain, A. K. (2010). Data clustering: 50 years beyond $k$-means. *Pattern Recognition Letters* 31, pp. 651–666.
50. Kohonen, T. (1990). The self-organizing map. *Proceedings of the IEEE* 78, pp. 1464–1480.
51. Pakhira, M. K., Bandyopadhyay, S. and Maulik, U. (2005). A study of some fuzzy cluster validity indices, genetic clustering and application to pixel classification. *Fuzzy Sets and Systems* 155, pp. 191–214.
52. Thimm, O. *et al.* (2004). MAPMAN: A user-driven tool to display genomics data sets onto diagrams of metabolic pathways and other biological processes. *The Plant Journal* 37, pp. 914–939.
53. Strehl, A. and Ghosh, J. (2002). Cluster ensembles — a knowledge reuse framework for combining partitions. *The Journal of Machine Learning Research* 3, pp. 583–617.
54. Charrad, M., Ghazzali, N., Boiteau, V. and Niknafs, A. (2014). NbClust: An R package for determining the relevant number of clusters in a data set. *Journal of Statistical Software* 61, pp. 1–35.
55. Anders, S. and Huber, W. (2010). Differential expression analysis for sequence count data. *Genome Biology*, 11, R106.

Chapter 7

# High-Performance Computing for Measurement of Cancer Gene Signatures

Hsueh-Ting Chu

*Department of Computer Science and Information Engineering,
Asia University, Taichung 41354, Taiwan*

## Abstract

For precision medicine (PM), the diagnosis and treatment of diseases get assistance from personalized genetic information such as risk alleles and gene expression profiles. Gene signatures are rapidly developing tools for the diagnosis and treatment of cancers. For the measurement of gene signatures, the computation of gene expression levels from RNA-Seq technology is an extremely time-consuming process since the massive amount of RNA-Seq data. Therefore, the acceleration of bioinformatics algorithms had been studied with high-performance computing frameworks, i.e. Apache Hadoop and Spark to process massive data sets in parallel. We illustrate the pipelines for RNA-Seq data processing for gene signatures and collect different HPC methods.

*Keywords*: bioinformatics, high-performance computing, gene signature, sequence analysis.

## 7.1. Introduction

In 1980s, DNA microarrays were invented to measure the expression levels of large numbers of genes simultaneously [1]. After that, identification of potential biomarkers from microarray experiments became a hot topic [2–4]. Different types of DNA microarrays were developed for the purpose, such as antibody microarray [5] and fusion genes microarray [6]. For the detection of different specific cancers, there were a lot of commercialized biochips (microarray). Table 7.1 lists some commercialized microarrays for early detection of diseases.

However, disease prediction with these biomarker arrays was sometime limited in terms of sensitivity and accuracy. In 1995, the technology of expression-based biomarkers obtained a leap by the improvement of microarray methods such as Serial Analysis of Gene Expression (SAGE) and a single or combined group of biomarkers were called a gene signature which represents a meaningful pattern of gene expression [7, 8]. In 2007, the gene signature Mammaprint was published by The Netherlands Cancer Institute (NKI) [9]. MammaPrint is a tool of prognostic test for early stage

Table 7.1.  Array-based biomarker.

| Array product | Targets |
|---|---|
| Gastric Cancer Human Biomarker Array | CA19-9, CA72-4, CEA, Pepsinogen1, and Pepsinogen 2 |
| Human Angiogenesis Array | IGF-1, IL-10, IL-12 p40, IL-12 p70, IL-17, IL-1 alpha, IL-1 beta, IL-2, IL-4, IL-6, IL-8, IP-10, I-TAC, Leptin, LIF, MCP-1, MCP-2, MCP-3, MCP-4, MMP-1, MMP-9, PDGF-BB, PECAM-1, PLGF, RANTES, TGF alpha, TGF beta 1, TGF beta 3, Tie-1, Tie-2, TIMP-1, TIMP-2, TNF alpha, TNF beta, TPO, uPAR, VEGF, VEGFR2, VEGF, R3, VEGF-D, and more |
| Human Cancer Biomarker Array | ACT, ACTH, Adipolean Variant, Adiponectin, AFP, Androgen Receptor, Ang-1, Ang-2, APLP2, ApoE3, ApoF, ApoL1, ApoL2, Beta-2-Microglobulin, beta-NGF, BDNF, CA15-3, CA19-9, CA125, Cadherin-pan, CEA, Collagen I, and more |
| Human Immune Checkpoint Molecule Array | B7-1/CD80, B7-2/CD86, B7-H1/PD-L1, B7-H2/ICOSL, B7-H3, CD28, CTLA-4, ICOS, PD-1, and PD-L2/B7-DC |

*Source*: https://www.biocat.com/proteomics/protein-expression-profiling/quantibody-arr-ays-glass-based-including-standards.

breast cancer patients to predict risk of metastasis using the expression of 70 genes. It classified breast cancer patients into two classes of high-risk and low-risk and it helps physicians to determine whether or not a patient will need chemotherapy. The evaluation of risk is according to the 5-year outcome of breast cancer operation. MammaPrint was the first IVDMIA (acronym of *In Vitro* Diagnostic Multivariate Index) Assays to be approved by the Food and Drug Administration (FDA). Following MammaPrint, different gene signatures for breast cancer were quickly developed, such as Oncotype DX [10], BluePrint [11, 12], TargetPrint [13, 14], and Theraprint [15]. Besides, there are gene signatures developed for other cancers, for example, Chen *et al.* developed a five-gene signature for non-small-cell lung cancer with the five genes DUSP6, MMD, STAT1, ERBB3, and LCK [16]. Table 7.2 lists a part of published gene signatures for colon cancers.

Microarray-based gene signatures have influenced the way of personized medicine. Gene signatures provides promising subtyping for a specific cancer. In recent years, Next-generation sequencing (NGS) technologies have revolutionized the study of genomics and molecular biology [17]. RNA sequencing with NGS methods (i.e. RNA-Seq) has the potential to bridge tumor genotypes and their phenotypic consequences such as cancer gene signatures [18]. Compared to microarrays, RNA-Seq technology can detect more differentially expressed genes with lower abundance [19]. Therefore, there were more and more projects using RNA-Seq technologies instead of microarray technologies for the last several years. For example, Agilent Technologies and Agendia are jointly developing an RNA-Seq kit version

Table 7.2.   Colon cancer gene signatures.

| Gene signatures | Targets |
| --- | --- |
| Coloprint | 18 genes |
| Oncotype DX<br>   Colon Cancer | 12 genes: Ki-67, C-MYC, MYBL2, FAP, BGN, INHBA,<br>   GADD45B with references (ATP5E, PGK1, GPX1, UBB,<br>   VDAC2) |
| ColoNext (NGS) | 14 genes: APC, BMPR1A, CDH1, CHEK2, EPCAM, MLH1,<br>   MSH2, MSH6, MUTYH, PMS2, PTEN, SMAD4, STK11,<br>   and TP53 |

*Source*: One-page summaries of the genetic tests for cancers https://www.ncbi.nlm.nih.gov/books/NBK285334/.

of Agendia's currently marketed MammaPrint and BluePrint tests. The new next-generation sequencing assays of the MammaPrint and BluePrint tests will afford patients to get better individualized treatment management.

However, modern RNA-Seq produced massive reads for a single run, and a typical read file is usually over 10 Gb. It demands a good computing environment to process the raw sequencing data. There had been many studied to utilize high-performance computing (HPC) resources to accelerating the pipeline of RNA-Seq analysis [20]. The Apache Hadoop open source project, which exploits the MapReduce framework and a distributed file system, has recently given bioinformatics researchers an opportunity to achieve scalable, efficient, and reliable computing performance with computer clusters or using cloud computing services such as AWS and GCP [21]. We discuss different studies for both gene-level and transcript-level computations on RNA-Seq data in this chapter.

## 7.2.  Background

### 7.2.1.  *RNA-Seq Experiment*

RNA sequencing (RNA-Seq) is revolutionizing the study of the transcriptome. Figure 7.1 shows the procedure of a typical RNA-Seq experiment. In general, the first step in RNA sequencing is the extraction of RNA and also removal of DNA. Then, primers are then added and will hybridize to complementary RNA sequences. In the presence of a reverse transcriptase and deoxynucleotide triphosphates (dNTPs) (A, T, G, C), the RNA insertions become the template for the synthesis of the first complementary strand of DNA (first strand synthesis). The DNA–RNA hybrids synthesized during first strand cDNA synthesis are converted into full-length double-stranded cDNAs. The primers for synthesis of the second strand cDNA are create by RNase. DNA Polymerase I extends the newly created 3′ strand, using the first-strand cDNA as a template. In this way, the remaining segments of mRNA in the cDNA–RNA hybrid are replaced with the newly synthesized second strand of cDNA.

After library preparation, the high-throughput sequencing by the Illumina platforms use the Sequencing-by-synthesis (SBS) technology. It uses four fluorescently-labeled nucleotides to sequence the tens of millions of

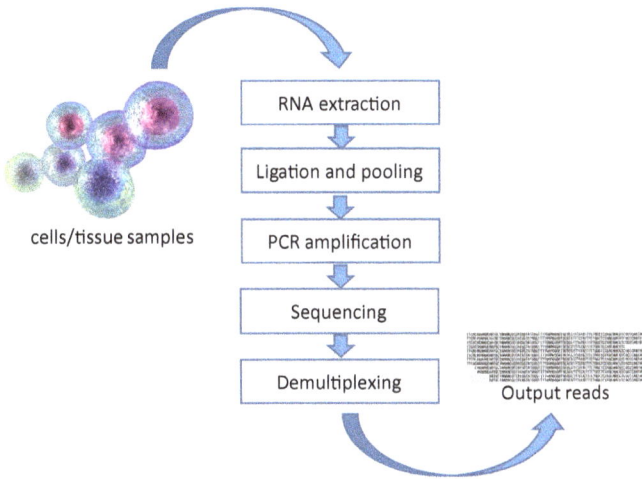

Fig. 7.1.    Steps of a RNA-Seq experiment.

clusters on the flow cell surface in parallel. During each sequencing cycle, a single labeled deoxynucleoside triphosphate (dNTP) is added to the nucleic acid chain. The nucleotide label serves as a terminator for polymerization, so after each dNTP incorporation, the fluorescent dye is imaged to identify the base and then enzymatically cleaved to allow incorporation of the next nucleotide. Since all four reversible terminator-bound dNTPs (A, C, T, G) are present as single, separate molecules, natural competition minimizes incorporation bias. Base calls are made directly from signal intensity measurements during each cycle. The RNA-Seq datasets using an Illumina sequencer usually have very lower error rates compared to other technologies.

### 7.2.2.  *Measurement of Cancer Gene Signatures with RNA-Seq*

The successful gene signatures including MammaPrint/BluePrint and other gene signatures ignited the need to develop NGS-based gene signatures recently since a single whole-transcriptome sequencing test can finish the expression measurement of all the genes involved in different gene signatures. Some studies are trying to move the MammaPrint breast cancer test from microarray to RNA-Seq for the measurement of the prognostic 70-gene profile [22]. The 70 genes in MammaPrint (in Table 7.3) includes

Table 7.3.    Breast cancer gene signatures.

| Gene signatures | Targets |
| --- | --- |
| MammaPrint | 70 Genes: AA404325, AA834945, AI224578, AI283268, ALDH4A1, AP2B1, AW014921, AYTL2, BBC3, C16orf61, C20orf46, C9orf30, CCNE2, CDC42BPA, CDCA7, CENPA, COL4A2, DCK, DIAPH3, DIAPH3.1, DIAPH3.2, DTL, EBF4, ECT2, EGLN1, ESM1, EXT1, FBXO31, FGF18, FLT1, GMPS, GNAZ, GPR126, GPR180, GSTM3, HRASLS, IGFBP5, IGFBP5, LGP2, LOC286052, LOC643008, MCM6, MELK, MMP9, MS4A7, MTDH, NDC80, NMU, NUSAP1, ORC6L, OXCT1, PALM2-AKAP2, ECI2, ECI2.1, PITRM1, PRC1, QSOX2, RAB6B, RFC4, RTN4RL1, RUNDC1, SCUBE2, SERF1A, SLC2A3, STK32B, TGFB3, TSPYL5, UCHL5, WISP1, ZNF533 |
| BluePrint | 80-Gene: ABAT, ABCC11, ACADSB, ACBD4, ADM, AFF3, AFF3, AGR2, AR, BCL2, BECN1, BTD, BTRC, CA12, CA12, CAPN13, CCDC74B, CDC25B, CDCA7, CELSR1, CELSR2, CHAD, CHAD, COQ7, DBNDD2, DHRS2, DNALI1, ELOVL5, ESR1, FOXA1, FOXC1, GATA3, GOLSYN, GREB1, HDAC11, HK3, HMGCL, IL6ST, IRS1, KIAA1370, KIAA1737, KIF20A, LILRB3, LRIG1, MAGED2, MLPH, MSN, MYB, MYO5C, NAT1, NPY1R, NUDT6, OCIAD1, PARD6B, PERLD1, PGR, PPAPDC2, PREX1, PRR15, REEP6, RERG, RTN4RL1, RUNDC1, S100A8, SCUBE2, SLC16A6, SOX11, SPEF1, SUSD3, TAPT1, TBC1D9, TCTN1, THSD4, TMC4, TMEM101, TMSB10, TPRG1, UBXD3, VAV3, XBP1 |
| TargetPrint | 3 genes: ER, PR, HER2 |
| Theraprint | 59 genes |
| Oncotype DX | 21 genes: ESR1, PGR, BCL2, SCUBE2, Ki67, STK15, Survivin, CCNB1, MYBL2, HER2, GRB7, MMP11, CTSL2, GSTM1, CD68, BACG1 with references (ACTB, GAPDH, RPLPO, GUS, TFRC) |
| BreastNext (NGS) | 14 genes: ATM, BARD1, BRIP1, CDH1, CHEK2,MRE11A, MUTYH, NBN, PALB2, PTEN, RAD50, RAD51C, STK11, and TP53 |

*Source*: Partial information cited from One-page summaries of the genetic tests for cancers
https://www.ncbi.nlm.nih.gov/books/NBK285334/.

three genes with isoforms including three isoforms of DIAPH3, two isoforms of IGFBP5 and two isoforms of ECI2. Therefore, it requires not only
gene-level assessment of expressions, but also transcript-level assessment
of expressions for the three genes (DIAPH3, IGFBP5, and ECI2). We list

Table 7.4.   Bioinformatics tools for measurement of cancer gene signatures with RNA-Seq.

| Bioinformatics tools | Functions |
| --- | --- |
| Tophat | RNA-Seq aligner |
| STAR | RNA-Seq aligner |
| HISAT2 | RNA-Seq aligner |
| Bowtie2 | RNA-Seq aligner |
| Cufflinks | Transcript assembly and quantification |
| StringTie | Transcript assembly |
| HTSeq | Expression assessment |
| Ballgown | Expression assessment |
| Kallisto | Expression assessment with Pseudoalignment |
| Salmon | Expression assessment with Pseudoalignment |

state-of-the-art Bioinformatics tools for the measurement of cancer gene signatures with RNA-Seq in Table 7.4.

For gene-level expression quantification, there are a lot of popular workflows, such as Tophat-HTSeq, Tophat-Cufflinks, STAR-HTSeq, Kallisto, and Salmon. The bioinformatics tools Tophat [23] and STAR [24] are two RNA-Seq aligners that map raw reads with a reference genome. And the tools HTSeq [25] and Cufflinks [26] then estimate the abundance of transcripts for expression assessment from the alignments by RNA-Seq aligners. Besides, Kallisto [27] and Salmon [28] are two tools for quantifying the expression of transcripts using so-called pseudoalignment methods. From a benchmarking study, it showed that the workflows Tophat-HTSeq and STAR-HTSeq have similar assessment results and slightly better than Tophat-Cufflinks [29]. The pseudoalignment methods Kallisto and Salmon even reach better assessment. However, every workflow produces a set of non-concordant expressions [29]. HISAT2 is a new RNA-Seq aligners which improve the alignment of reads by the information of pseudogenes and SNPs [30].

For transcript-level expression quantification, the pseudoalignment methods Kallisto and Salmon can be directly use for quantification of transcript-level expression. Besides, RSEM is another popular tool which can run with the aligners Bowtie2 [31] or STAR. There is a new workflow of transcript-level analysis proposed recently with HISAT2 and two other

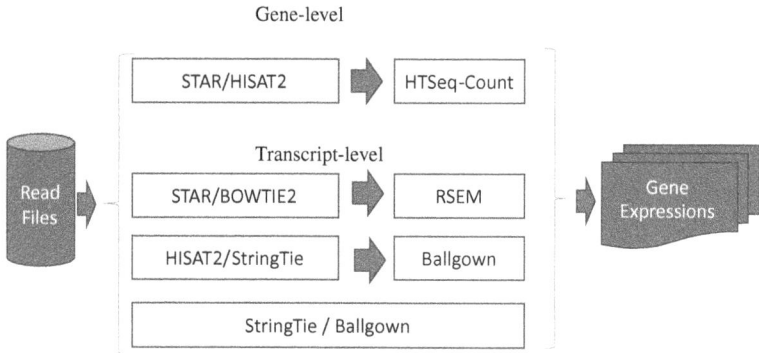

Fig. 7.2.  Workflows of gene-level and transcript-level expression measurement with RNA-Seq.

tools StringTie and Ballgown [32]. Figure 7.2 shows popular workflows of expression quantification with RNA-Seq.

### 7.2.3.  *High-performance Computing (HPC) with Apache Hadoop and Spark*

For HPC, there are many open-source software framework including Apache Hadoop, Apache Spark, Apache Hama, Apache Avro, Docker Swarm, etc. Most of these software framework can be mixed together for achieving the aim of HPC. The acceleration of massive computations usually combines the capability of data and process migrations between a grid or a cloud service. Apache Hadoop contains three components to achieve parallel processing of computations: YARN, HDFS, and MapReduce. (1) Hadoop YARN is to split up the functionalities of resource management and job scheduling/monitoring into separate daemons composed of ResourceManager, ApplicationMaster, NodeManager, and YarnClients. (2) The Hadoop Distributed File System (HDFS) is a distributed file system designed to run on commodity hardware. An HDFS instance may consist of hundreds or thousands of server machines, each storing part of the file system's data. (3) Hadoop MapReduce is a software framework for running applications in-parallel on large clusters. MapReduce jobs usually are processed by both the map tasks and the reduce tasks.

## 7.3.   RNA-Seq Analysis on HPC for Gene Signature

Figure 7.3 shows a conceptual workflow of computational analysis for RNA-Seq data with Apache Hadoop. The read files have to be split into independent chunks which are processed by the map tasks in a completely parallel manner. The framework sorts the outputs of the maps, which are then input to the reduce tasks. The framework takes care of scheduling tasks, monitoring them and deals with the failed tasks [21]. There had been many studies made for using HPC for RNA-Seq analysis shown in Table 7.5.

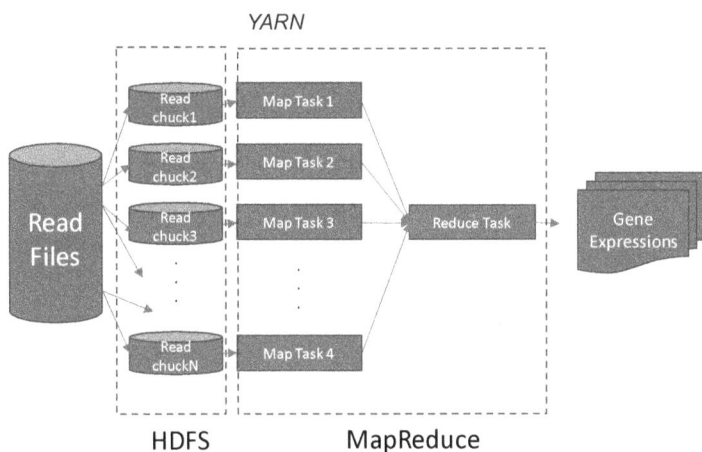

Fig. 7.3.   Basic functionality of high-performance computing for gene expression analysis.

Table 7.5.   HPC tools for gene expression analysis from RNA-Seq data.

| HPC tools | Functions |
| --- | --- |
| CloudBurst | HPC short-read aligner [33] |
| CloudAligner | HPC short-read aligner [34] |
| DistMap | HPC short-read aligner [35] |
| HSRA | HPC RNA-Seq aligner [36] |
| Hadoop-BAM | BAM utility for Hadoop [37] |
| ADAM | Genomics analysis APIs on HPC platform [38] |
| Halvade-RNA | Variant call analysis on HPC platform [39] |
| aRNApipe | HPC RNA-seq pipeline [20] |
| Myrna | Expression assessment with Pseudoalignment [40] |

## 7.4.   Conclusion

Most of the HPC tools in Table 7.5 can be roughly divided into four classes. The first class is the design of distributed mapping using Apache Hadoop or Spark framework. The second class is the adapter of traditional mapping tools on HPC platform. The third class is providing an interface mechanism of HPC for reinventing current Bioinformatics tools. The four class are the integration of Hadoop and R programs. However, most of the HPC tools are not considered by people because the bottleneck of dividing read files into HDFS chucks limit the improvement of performance. For example, HSRA is a Hadoop-based read aligner which is add the HDFS functions to the state-of-the-art HISAT2 program. It can boost the performance of a RNA-Seq mapping job on average 2.3 times with a 16-node multi-core cluster. On the contrary, each node can only provide 1/8 productivity of single computing node. Since the huge scale of sequencing data, it is still difficult to integrate the HPC tools for the measurement of cancer gene signatures. However, the success of the pseudoalignment aligners such as Kallisto and Salmon has broken the traditional framework of mapping-based RNA-Seq analysis. New HPC tools using pseudoalignment will be promising to accelerate the measurement of cancer gene signatures with RNA-Seq in the near future.

### References

1.  Taub, F. E., DeLeo, J. M. and Thompson, E. B. (1983). Sequential comparative hybridizations analyzed by computerized image processing can identify and quantitate regulated RNAs. *DNA* 2(4), pp. 309–327. doi: 10.1089/dna.1983.2.309.
2.  Sanchez-Pena, M. L., Isaza, C. E., Perez-Morales, J., Rodriguez-Padilla, C., Castro, J. M. and Cabrera-Rios, M. (2013). Identification of potential biomarkers from microarray experiments using multiple criteria optimization. *Cancer Medicine* 2(2), pp. 253–265. Epub 2013/05/02. doi: 10.1002/cam4.69. PubMed PMID: 23634293; PubMed Central PMCID: PMCPMC3639664.
3.  Sánchez-Peña, M. L., Isaza, C. E., Pérez-Morales, J., Rodríguez-Padilla, C., Castro, J. M. and Cabrera-Ríos, M. (2013). Identification of potential biomarkers from microarray experiments using multiple criteria optimization. *Cancer Medicine* 2(2), pp. 253–265. doi: 10.1002/cam4.69. PubMed PMID: PMC3639664.
4.  Tong, W., Ye, F., He, L., Cui, L., Cui, M., Hu, Y. *et al.* (2016). Serum biomarker panels for diagnosis of gastric cancer. *OncoTargets and Therapy* 9, pp. 2455–2463. doi: 10.2147/OTT.S86139. PubMed PMID: PMC4853138.

5. Zhu, Z. Q., Tang, J. S., Gang, D., Wang, M. X., Wang, J. Q., Lei, Z. *et al.* (2015). Antibody microarray profiling of osteosarcoma cell serum for identifying potential biomarkers. *Molecular Medicine Reports* 12(1), pp. 1157–1162. Epub 2015/03/31. doi: 10.3892/mmr.2015.3535. PubMed PMID: 25815525.

6. Lovf, M., Thomassen, G. O., Bakken, A. C., Celestino, R., Fioretos, T., Lind, G. E. *et al.* (2011). Fusion gene microarray reveals cancer type-specificity among fusion genes. *Genes, Chromosomes & Cancer* 50(5), pp. 348–357. Epub 2011/02/10. doi: 10.1002/gcc.20860. PubMed PMID: 21305644.

7. Schena, M., Shalon, D., Davis, R. W. and Brown, P. O. (1995). Quantitative monitoring of gene expression patterns with a complementary DNA microarray. *Science* 270(5235), p. 467.

8. Velculescu, V. E., Zhang, L., Vogelstein, B. and Kinzler, K. W. (1995). Serial analysis of gene expression. *Science* 270(5235), p. 484.

9. Tian, S., Roepman, P., van't Veer, L. J., Bernards, R., de Snoo, F. and Glas, A. M. (2010). Biological functions of the genes in the Mammaprint breast cancer profile reflect the hallmarks of cancer. *Biomarker Insights* 5, pp. 129–138. doi: 10.4137/BMI.S6184. PubMed PMID: PMC2999994.

10. Toole, M. J., Kidwell, K. M. and Van Poznak, C. (2014). Oncotype dx results in multiple primary breast cancers. *Breast Cancer: Basic and Clinical Research* 8, pp. 1–6. Epub 2014/01/24. doi: 10.4137/bcbcr.S13727. PubMed PMID: 24453493; PubMed Central PMCID: PMCPMC3891573.

11. Whitworth, P., Stork-Sloots, L., de Snoo, F. A., Richards, P., Rotkis, M., Beatty, J. *et al.* (2014). Chemosensitivity predicted by bluePrint 80-gene functional subtype and MammaPrint in the prospective neoadjuvant breast registry symphony trial (NBRST). *Annals of Surgical Oncology* 21(10), pp. 3261–3267. doi: 10.1245/s10434-014-3908-y. PubMed PMID: PMC4161926.

12. Krijgsman, O., Roepman, P., Zwart, W., Carroll, J. S., Tian, S., de Snoo, F. A. *et al.* (2012). A diagnostic gene profile for molecular subtyping of breast cancer associated with treatment response. *Breast Cancer Research and Treatment* 133(1), pp. 37–47. doi: 10.1007/s10549-011-1683-z.

13. Viale, G., Slaets, L., Bogaerts, J., Rutgers, E., van't Veer, L., Piccart-Gebhart, M. J. *et al.* (2014). High concordance of protein (by IHC), gene (by FISH; HER2 only), and microarray readout (by TargetPrint) of ER, PgR, and HER2: results from the EORTC 10041/BIG 03-04 MINDACT trial. *Annals of Oncology* 25(4), pp. 816–823. doi: 10.1093/annonc/mdu026. PubMed PMID: PMC3969556.

14. Roepman, P., Horlings, H. M., Krijgsman, O., Kok, M., Bueno-de-Mesquita, J. M., Bender, R. *et al.* (2009). Microarray-based determination of estrogen receptor, progesterone receptor, and HER2 receptor status in breast cancer. *Clinical Cancer Research* 15(22), pp. 7003–11. Epub 2009/11/06. doi: 10.1158/1078-0432.Ccr-09-0449. PubMed PMID: 19887485.

15. Witkiewicz, A. K., Balaji, U. and Knudsen, E. S. (2014). Systematically defining single-gene determinants of response to neoadjuvant chemotherapy reveals specific biomarkers. *Clinical Cancer Research: An Official Journal of the American Association for Cancer Research* 20(18), pp. 4837–4848. doi: 10.1158/1078-0432.CCR-14-0885. PubMed PMID: PMC5286972.

16.  Chen, H.-Y., Yu, S.-L., Chen, C.-H., Chang, G.-C., Chen, C.-Y., Yuan, A. *et al.* (2007). A five-gene signature and clinical outcome in non–small-cell lung cancer. *The New England Journal of Medicine* 356(1), pp. 11–20. doi: 10.1056/NEJMoa060096.

17.  Mardis, E. R. (2008). The impact of next-generation sequencing technology on genetics. *Trends in Genetics* 24(3), pp. 133–141. doi: https://doi.org/10.1016/j.tig.2007.12.007.

18.  Cieślik, M. and Chinnaiyan, A. M. (2017). Cancer transcriptome profiling at the juncture of clinical translation. *Nature Reviews Genetics* 19, p. 93. doi: 10.1038/nrg.2017.96.

19.  Zhao, S., Fung-Leung, W.-P., Bittner, A., Ngo, K. and Liu, X. (2014). Comparison of RNA-Seq and microarray in transcriptome profiling of activated T cells. *PLoS ONE* 9(1), p. e78644. doi: 10.1371/journal.pone.0078644.

20.  Alonso, A., Lasseigne, B. N., Williams, K., Nielsen, J., Ramaker, R. C., Hardigan, A. A. *et al.* (2017). aRNApipe: A balanced, efficient and distributed pipeline for processing RNA-seq data in high-performance computing environments. *Bioinformatics* 33(11), pp. 1727–1729. doi: 10.1093/bioinformatics/btx023.

21.  Zou, Q., Li, X.-B., Jiang, W.-R., Lin, Z.-Y., Li, G.-L. and Chen, K. (2014). Survey of MapReduce frame operation in bioinformatics. *Briefings in Bioinformatics* 15(4), pp. 637–647. doi: 10.1093/bib/bbs088.

22.  Mittempergher, L., Spangler, J. B., Snel, M. H., Delahaye, L. J., Rink, I. D., Tian, S. *et al.* (2017). Abstract 5409: Assessment of the MammaPrint 70-gene profile using RNA sequencing technology. *Cancer Research* 77(13 Supplement), p. 5409.

23.  Trapnell, C., Pachter, L. and Salzberg, S. L. (2009). TopHat: Discovering splice junctions with RNA-Seq. *Bioinformatics* 25(9), 1105–1111. Epub 2009/03/18. doi: 10.1093/bioinformatics/btp120. PubMed PMID: 19289445; PubMed Central PMCID: PMCPMC2672628.

24.  Dobin, A., Davis, C. A., Schlesinger, F., Drenkow, J., Zaleski, C., Jha, S. *et al.* (2013). STAR: ultrafast universal RNA-seq aligner. *Bioinformatics* 29(1), 15–21. doi: 10.1093/bioinformatics/bts635. PubMed PMID: PMC3530905.

25.  Anders, S., Pyl, P. T. and Huber, W. (2015). HTSeq — a Python framework to work with high-throughput sequencing data. *Bioinformatics*. 31(2), pp. 166–169. Epub 2014/09/28. doi: 10.1093/bioinformatics/btu638. PubMed PMID: 25260700; PubMed Central PMCID: PMCPMC4287950.

26.  Ghosh, S. and Chan, C. K. (2016). Analysis of RNA-Seq data using TopHat and cufflinks. *Methods in Molecular Biology (Clifton, NJ)*. 1374, pp. 339–361. Epub 2015/11/01. doi: 10.1007/978-1-4939-3167-5_18. PubMed PMID: 26519415.

27.  Bray, N. L., Pimentel, H., Melsted, P. and Pachter, L. (2016). Near-optimal probabilistic RNA-seq quantification. *Nature Biotechnology* 34(5), pp. 525–527. Epub 2016/04/05. doi: 10.1038/nbt.3519. PubMed PMID: 27043002.

28.  Patro, R., Duggal, G., Love, M. I., Irizarry, R. A. and Kingsford, C. (2017). Salmon provides fast and bias-aware quantification of transcript expression. *Nature Methods* 14(4), pp. 417–419. Epub 2017/03/07. doi: 10.1038/nmeth.4197. PubMed PMID: 28263959; PubMed Central PMCID: PMCPMC5600148.

29. Everaert, C., Luypaert, M., Maag, J. L. V., Cheng, Q. X., Dinger, M. E., Hellemans, J. *et al.* (2017). Benchmarking of RNA-sequencing analysis workflows using whole-transcriptome RT-qPCR expression data. *Scientific Reports* 7(1), p. 1559. doi: 10.1038/s41598-017-01617-3.
30. Kim, D., Langmead, B. and Salzberg, S. L. (2015). HISAT: A fast spliced aligner with low memory requirements. *Nature Methods* 12(4), pp. 357–360. Epub 2015/03/10. doi: 10.1038/nmeth.3317. PubMed PMID: 25751142; PubMed Central PMCID: PMCPMC4655817.
31. Langmead, B. and Salzberg, S. L. (2012). Fast gapped-read alignment with Bowtie 2. *Nature Methods* 9(4), pp. 357–359. Epub 2012/03/06. doi: 10.1038/nmeth.1923. PubMed PMID: 22388286; PubMed Central PMCID: PMCPMC3322381.
32. Pertea, M., Kim, D., Pertea, G. M., Leek, J. T. and Salzberg, S. L. (2016). Transcript-level expression analysis of RNA-seq experiments with HISAT, StringTie and Ballgown. *Nature Protocols* 11(9), pp. 1650–1667. Epub 2016/08/26. doi: 10.1038/nprot.2016.095. PubMed PMID: 27560171; PubMed Central PMCID: PMCPMC5032908.
33. Schatz, M. C. (2009). CloudBurst: Highly sensitive read mapping with MapReduce. *Bioinformatics* 25(11), pp. 1363–1369. doi: 10.1093/bioinformatics/btp236.
34. Nguyen, T., Shi, W. and Ruden, D. (2011). CloudAligner: A fast and full-featured MapReduce based tool for sequence mapping. *BMC Research Notes* 4(1), p. 171. doi: 10.1186/1756-0500-4-171.
35. Pandey, R. V. and Schlötterer, C. (2013). DistMap: A toolkit for distributed short read mapping on a hadoop cluster. *PLoS ONE* 8(8), p. e72614. doi: 10.1371/journal.pone.0072614. PubMed PMID: PMC3751911.
36. Exposito, R. R., Gonzalez-Dominguez, J. and Tourino, J. (2018). HSRA: Hadoop-based spliced read aligner for RNA sequencing data. *PLoS One* 13(7), p. e0201483. Epub 2018/08/01. doi: 10.1371/journal.pone.0201483. PubMed PMID: 30063721; PubMed Central PMCID: PMCPMC6067734.
37. Niemenmaa, M., Kallio, A., Schumacher, A., Klemelä, P., Korpelainen, E. and Heljanko, K. (2012). Hadoop-BAM: directly manipulating next generation sequencing data in the cloud. *Bioinformatics* 28(6), pp. 876–877. doi: 10.1093/bioinformatics/bts054. PubMed PMID: PMC3307120.
38. Nothaft, F. A., Massie, M., Danford, T., Zhang, Z., Laserson, U., Yeksigian, C. *et al.* (2015). Rethinking Data-Intensive Science Using Scalable Analytics Systems. *Proceedings of the 2015 ACM SIGMOD International Conference on Management of Data*, Melbourne, Victoria, Australia. 2742787: ACM, pp. 631–646.
39. Decap, D., Reumers, J., Herzeel, C., Costanza, P. and Fostier, J. (2017). Halvade-RNA: Parallel variant calling from transcriptomic data using MapReduce. *PLoS ONE* 12(3), p. e0174575. doi: 10.1371/journal.pone.0174575.
40. Langmead, B., Hansen, K. D. and Leek, J. T. (2010). Cloud-scale RNA-sequencing differential expression analysis with Myrna. *Genome Biology* 11(8), p. R83. doi: 10.1186/gb-2010-11-8-r83.

Chapter 8

# High-Performance Computing in Tandem Mass Spectrometry (MS/MS) Data Processing

Li Chuang and Lin Feng

*School of Computer Science and Engineering,*
*Nanyang Technological University,*
*Singapore 639798*

## Abstract

Tandem mass spectrometry (MS/MS) data have become widespread, and the exploitation of this technology has gone from laboratories to general users in industry. Processing the huge amount of MS/MS data turns out to be a major computational challenge. In the era of big data mining, faster transformation of MS/MS data into effective information has been one of the most significant challenges in bioinformatics. Meanwhile, high-performance computing (HPC) has advanced rapidly in various fields. Accordingly, application of HPC in bioinformatics to gain insight from MS/MS data efficiently has been emphasized in both industry and academia. In this work, we review HPC in MS/MS data processing, present our research work, and analyze the relevant work in this domain. To provide a useful and comprehensive perspective, we categorize the research by both the HPC in database search sequence analysis and the HPC in *de novo* sequence analysis.

*Keywords*: bioinformatics, high-performance computing (HPC), MS/MS, sequence analysis.

## 8.1. Introduction

Discovery for the potential, non-obvious, useful information from large quantities of data has become increasingly important in various domains [1], and bioinformatics is no exception. Extreme volumes of biomedical data including image, signal data, and omics [2–6] have been accumulated, and the potential for applications in healthcare research and biological has caught the attention of both academia and industry [7, 8].

In recent years, tandem mass spectrometry (MS/MS) has become the most significant tool of choice for the analysis proteins in high-throughput proteomics studies [9–12]. Benefiting from the advances in the mass spectrometry technology, modern mass spectrometers are now capable of producing hundreds of thousands of MS/MS spectra per experiment. For example, the data generated from the research on human protein expression profiling in 2017 exceeds 5.1 TB; this can also be confirmed by the huge increase in mass spectra data recorded by PRIDE. The extremely large amounts of data have brought difficulties to interpret these MS/MS spectra data in peptide matches [13]. Developing efficient MS/MS data processing method is an urgent need for MS/MS data processing in computational proteomics.

To address the aforementioned needs, a lot of research efforts have been made toward the incorporation of high-performance computing (HPC) into biological and biomedical data processing [14–19]. Relevant examples include network and systems biology studies [14, 15], medical imaging [16, 17], sequence analysis [18, 19] such as proteomic sequencing and genomic sequencing.

The idea of using many computers [20, 21], clustered together to work as a computational cluster has inspired many developments based on exploiting multi-processor systems [22–25]. The main purposes of the HPC technology design are to utilize highly heterogeneous computing resources [26, 27]. In this regard, distributed networks can realize a very high level of aggregate performance in compute-intensive bioinformatics applications [28]. Meanwhile, the emergence of specialized hardware devices such as graphic processing units (GPUs), field programmable gate arrays (FPGAs), and Many Integrated Core (MIC) exhibit the potential to real-time analysis of

MS/MS data. The increasing computational demands of MS/MS data analysis applications can now benefit from these compact hardware components, taking advantage of the small size and relatively low cost of these units as compared to clusters or networks of computers. These aspects are of great importance in the definition of sequence analysis, in which the payload is an important parameter.

HPC comprises a set of integrated acceleration hardware and programming system which can greatly assist in the task of solving large-scale data processing problems. With the rapid development of HPC technology, many bioinformatics applications based on HPC have been developed and successfully implemented in reality, such as sequence analysis [29, 30], gene and protein expression [31], and analysis of cellular organization [32]. Of these, the sequence analysis is academia, and industry focused their attention on an important research topic in one. The goal of this work is to conduct some preliminary investigations along this direction.

In this work, we focus on recent advances in the field of HPC applied to MS/MS data analysis. The remainder of the chapter is organized following the general order of increasing system building block size in HPC. Specifically, Section 8.2 gives a background on the tandem mass spectrometry, sequence analysis, and HPC. In Section 8.3, we focus on the HPC in MS/MS data processing, with emphasis HPC in peptide database-search sequencing and *de novo* peptide sequencing. Finally, in Section 8.4, we conclude this study and discuss the future directions of research.

## 8.2. Background

### 8.2.1. *Tandem Mass Spectrometry (MS/MS)*

The goal of proteomics is to identify and characterize all proteins. Tandem mass spectrometry (MS/MS), coupled with liquid chromatography (LC) has become the most significant tool of choice for the analysis proteins in high-throughput proteomics studies. It has been particularly useful for the identification, characterization, and comparative analysis of protein mixtures, which composed of thousands of different proteins. Figure 8.1 shows an example of MS/MS spectrum, which contains the measured m/z and intensity of the fragments, represented by the peaks. Different ionization

| 88 | 145 | 292 | 405 | 534 | 663 | 778 | 907 | 1020 | 1166 | b ions |
|----|-----|-----|-----|-----|-----|-----|-----|------|------|--------|
| S | G | F | L | E | E | D | E | L | K | |
| 1166 | 1080 | 1022 | 875 | 762 | 633 | 504 | 389 | 260 | 146 | y ions |

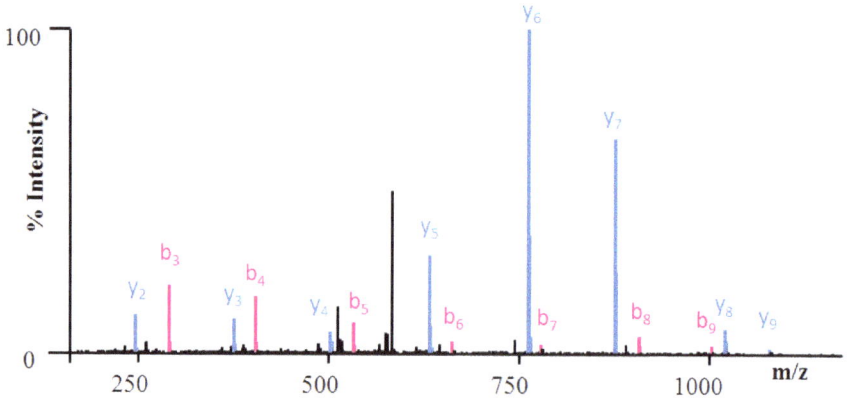

Fig. 8.1.   An example of MS/MS spectrum.

# Peptide Fragmentation

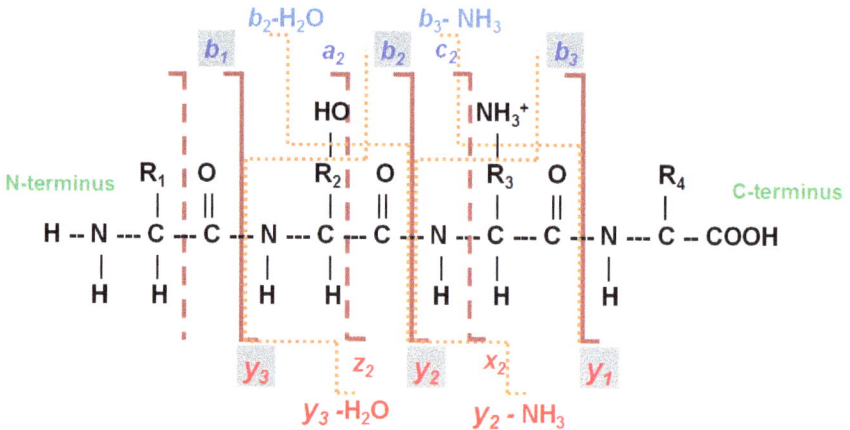

Fig. 8.2.   The type fragments.

methods have dramatic impact on the propensities for producing particular fragment ion types. For example, in CID, there are six series of fragment ions, which denote by type fragments C-terminal $x$, $y$ and $z$ and N-terminal $a$, $b$ and $c$ type fragments, as show as Fig. 8.2.

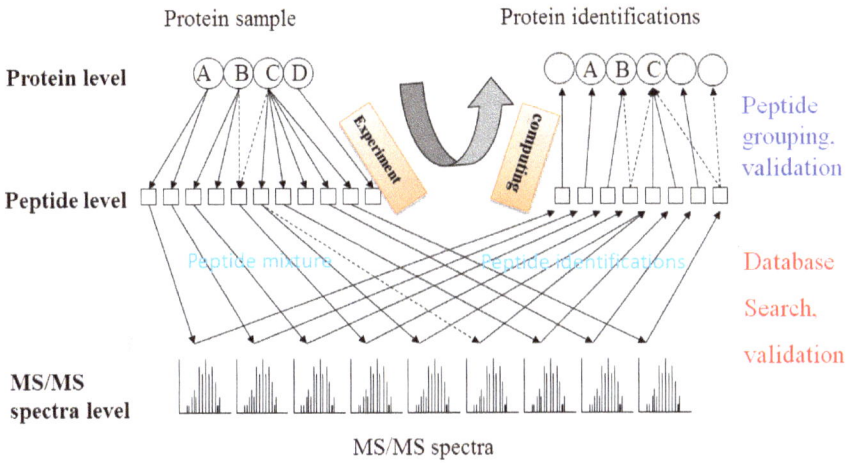

Fig. 8.3. The process of high-throughput mass spectrometry-based protein identification method.

A detailed process of the MS/MS-based approach to study complex protein mixtures is illustrated in Fig. 8.3.

First, because the intact proteins are not amenable for mass spectrometric identification, sample proteins are digestion cleaved into peptides by the enzyme trypsin. In the protein digestion, we can use strong cation exchange chromatography to reduce the complexity of the peptide mixture. Secondly, the peptide mixture is subjected to the liquid chromatography–tandem mass spectrometry. Finally, peptides are ionized and selected ions subjected to fragmentation in the collision cell to produce MS/MS. For each MS/MS, we must use computational methods to infer the peptides and proteins.

### 8.2.2. *MS/MS-based Sequence Analysis*

Computational methods have been used in biology for sequence analysis. High-throughput MS/MS-based shotgun method is the most effective and commonly used computational method for identifying peptides in a proteomics research. The procedure of shotgun methods [33] consists of two main stages: the experimental stage and the sequence analysis, as shown in Fig. 8.4 [34, 35].

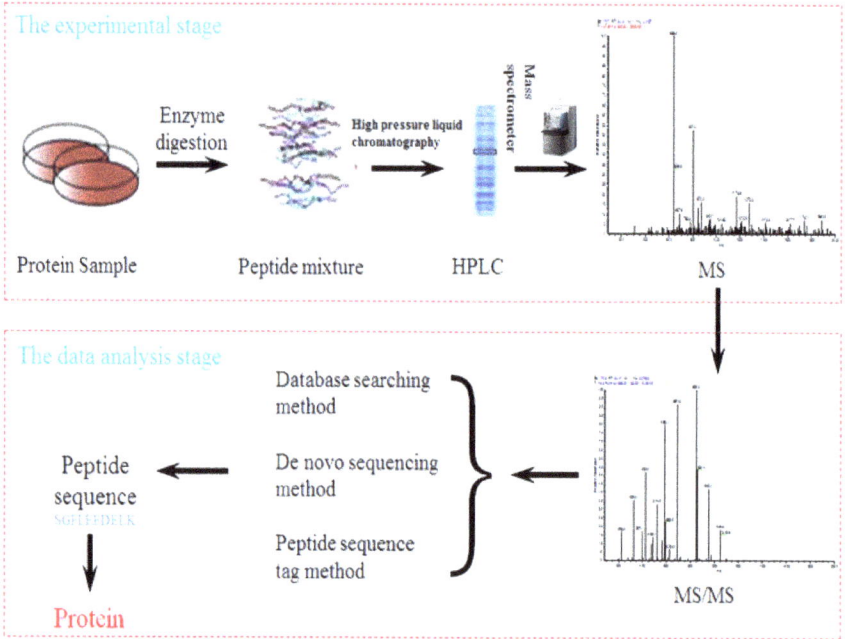

Fig. 8.4. Workflow of shotgun method.

Among them, the sequence analysis stage is the heart of correctly identified peptide and protein. The database search method [36] is the most widely used method in the sequence analysis phase and the key point of large-scale analysis of the protein sequence and PTMs with high sensitivity, accuracy, and throughput [33, 36]. Database search programs such as SEQUEST [37], Mascot [38], and X!!Tandem [39] are used to identify peptides to MS/MS spectra. The major limitation to this method is that it is highly dependent on the database. The other method is *de novo* peptide sequencing [40], which can directly extract a peptide sequence from a MS/MS spectrum, e.g. PEAKS [41] and UniNovo [42], etc. In recent years, *de novo* peptide sequencing has been widely acknowledged as an essential method for the discovery of new proteins and post-translational modifications (PTMs).

The extremely large amounts of data have brought difficulties to MS/MS sequence analysis. Moreover, noise and resulting from the presence of other impurities in the peptide mixture will produce interfering peaks, and the theory behind fragmentation is incomplete. Both of these problems can lead to very low sequencing efficiency and extremely misidentification or the inability to decide among numerous possible identifications.

### 8.2.3. *High-Performance Computing (HPC)*

Scientific computing usually includes the construction of numerical solution techniques and mathematical models to solve engineering, scientific and social scientific problems. In large-scale experiments, these models often require a huge number of computing resources to complete the job with care and efficiency. In general, these needs have been addressed by using HPC solutions and installed facilities such as clusters and super computers, which are the best solution available. Especially in recent years, the rapid development of HPC technologies provides the opportunity to greatly reduce the computational time of many scientific computing programs.

In the era of big data, new HPC architecture requires heterogeneous architectures by increasing the number of cores, co-processors, and accelerators. To improve computational performance of existing HPC, supplemental processors including Field Programmable Gate Arrays (FPGAs), Graphical Processing Units (GPUs), and now Intel MIC cards are being applied industrywide for this purpose. GPUs and Intel MIC was the most representative HPC technologies in recent years.

The designs of GPU are ideal for a particular class of applications with the following characteristics: computational requirements are large, parallelism is substantial, and throughput is more important than latency. Over the past several years, a growing field has identified other applications with similar characteristics and successfully implemented these applications onto the GPU.

Intel Many Integrated Core (MIC) Architecture is a many-core coprocessor (Intel Xeon Phi coprocessor) intended for highly parallel multithreaded application with high memory bandwidth needs, which focus on performances and power efficiency. It is based on an X86 Pentium core architecture but contains 512-bit-wide vector units and each coprocessor features

60+ cores clocked at 1 GHz or more, supporting 64-bit x86 instructions. These in-order cores support four ways hyper-threading, resulting in more than 240 logical cores. In principle, one great benefit of using Intel MIC technology, compared with other accelerators and coprocessors, is the simplicity of the development. Programmers do not have to learn a new programming language but may compile their source codes specifying MIC as the target architecture. The Traditional programming languages used for HPC, as Fortran/C/C++ and the parallel paradigms — MPI or OpenMP — can be directly employed regarding MIC. Typically MIC supports three typical kinds of programming models that can be used to design and implement parallel algorithms on MIC-based heterogeneous architectures. In most experiments, researcher have used the offload model to design and implement parallel algorithms which can make full use of the compute power of both the multi-core CPU and the many-core Xeon Phi hardware.

## 8.3.  MS/MS Data Analysis on HPC

Computational methods have been used in biology for sequence analysis. Due to their excessive time and memory overheads, existing sequence analysis tools suffer from low efficiency in processing this large amount of MS/MS spectra. More efficient sequence analysis tool is of paramount importance for computational biology researchers.

### 8.3.1.  *Database-search Sequence Analysis*

MS/MS-based database searching is a widely acknowledged and adopted method that assigns peptide sequence in shotgun proteomics. However, due to the rapid growth of spectra data produced in advanced mass spectrometer and more and more modified and digested peptides need to be considered in recent years, the overhead of large-scale data processing using database searching has become unacceptable. MR-Tandem is an efficient tool for parallel database search sequencing using Hadoop MapReduce on Amazon Web Services, which is developed by Pratt [43]. It adapts the widely used

X!Tandem peptide search engine [44]. Compare with other database-search methods, MR-Tandem offered several advantages:

1. Better correctness: MR-Tandem is the first to exploit the fault tolerance and scalability of Hadoop to create large-scale compute clusters where MPI versions may fail.
2. Excellent parallel performance: The search speedup of MR-Tandem 20% more than X!!Tandem [45].
3. Scalability: MR-Tandem scales is superior to other X!Tandem parallel implementations that search against the same proteins on all nodes.
4. Ease to use: MR-Tandem can be run in almost exactly the same way as the sequential versions.

MR-Tandem is an efficient parallelized version of X!Tandem that demonstrates excellent speedup. In addition, it is substantially the same as the original code, is run in the same manner, and produces identical output. The improved performance of MR-Tandem should be of benefit to researchers analyzing mass spectrometry data. The source code for MR-Tandem has been made freely available via http://sashimi.svn.sourceforge.net/viewvc/sashimi/trunk/trans_proteomic_pipeline/extern/xtandem/.

### 8.3.2. *De novo Sequence Analysis*

*De novo* peptide sequencing will play a very important role in MS/MS data processing as well. More importantly, *de novo* peptide sequencing can directly extract a peptide sequence from a MS/MS spectrum. As a result, *de novo* sequencing has been widely applied in the MS/MS data analysis. It was employed to improve the sensitivity and accuracy of database search by validating the results of database search and also to accelerate database search by taking *de novo* peptide sequence tags as a filter. Nevertheless, the relatively slow speed of *de novo* peptide sequencing tool of today has lessened the significance of assisting database searches. The current algorithms for *de novo* peptide sequencing are limited to small-scale datasets, which prohibits thorough analysis of large model in a short period of time.

Therefore, it is of great importance to improve the speed of *de novo* peptide sequencing [1, 2].

To address this problem, we present a high-performance Hadoop-based tool named MRUniNovo which parallelizes UniNovo [46] — a popular universal *de novo* sequencing tool to accelerate *de novo* peptide sequencing. MRUniNovo aims at increasing the performance and scalability of *de novo* peptide sequencing by using HPC technology without sacrificing correctness and accuracy in the result. To our best knowledge, MRUniNovo is the first tool for parallel *de novo* peptide sequencing.

Specifically, we proposed an optimized MapReduce framework, separated the process of *de novo* peptide sequencing into two phases, and de-signed the novel MRUniNovo. The workflow of MRUniNovo consists of two phases, as show in Fig. 8.5.

The first one is the sequential phase, where MS/MS spectra datasets are preprocessed. In this phase, MRUniNovo partitions a MS/MS spectra dataset into properly-sized chunks. The chunk size impacts on the performance of MRUniNovo, which is illustrated in the supplementary Fig. 8.6. As shown, when the chunk size was approximately 10MB, MRUniNovo achieved the best performance, indicated by the least computational time. Figure 8.6 also demonstrates when the chunk size exceeded 25MB, the computational time of MRUniNovo increased considerably because the Java VM was out of memory.

The second phase is the parallel part, where two types of tasks are executed in parallel, the Map task and Reduce task. A Map task scores candidate peptides and maps the results to key-value pairs in the form of <peptide; score>. A Reduce task summarizes the <peptide; score> pairs and identifies the peptides with the highest scores as the final results. Benefit from Hadoop Distributed File System (HDFS), MRUniNovo adopts an efficient data rebalancing scheme to automatically move MS/MS spectra data from overutilized DataNodes to underutilized DataNodes. Therefore, MRUninovo achieves high fault tolerance and maximizes its overall processing capacity. Moreover, MRUninovo guarantees the correctness and accuracy of the results by detects and recovers from runtime faults quickly and automatically. This is a pattern that parallelizes *de novo* peptide

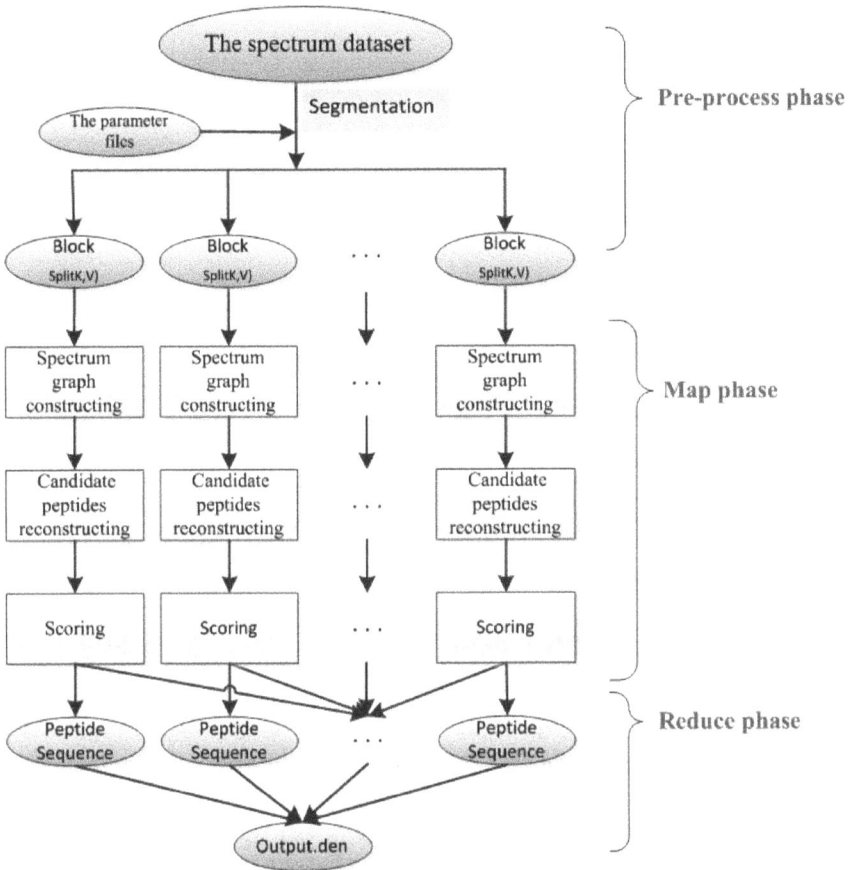

Fig. 8.5. Workflow of MRUniNovo.

sequencing by MapReduce. It also can be applicable to other *de novo* peptide sequencing.

MRUniNovo is a Java program designed to accelerate *de novo* peptide sequencing. We assume that readers are familiar with Java to some extent and are capable of running Java programs. It is available from https://github.com/Logic09/MRUniNovo1/tree/41a2148f267274a21bcddd b7869a524a53095023. Hardware requirements: Hadoop cluster. Each machine in the Hadoop cluster has a dual-core 3.2 GHz Intel Xeon processor, 500 GB Hard Disk Drive, JDK1.6, and Hadoop 2.2.0 installed, and

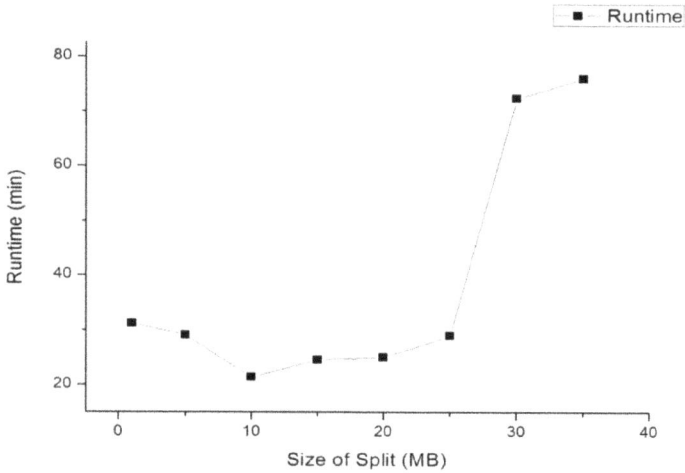

Fig. 8.6.   The effect of chunk size on the speedup of MRUniNovo.

requires a minimum of 2 GB RAM. The dataset is Fetal_Brain_bRP_Elite, which a MS/MS spectra dataset with high accuracy and high resolution.

Figure 8.7 shows the speedup of MRUniNovo against the number of machines in the Hadoop cluster. The performance of MRUniNovo increases with the size of the cluster. MRUniNovo gets more significant as the cluster size increases. Moreover, it is possible to reduce the processing time further by adding more Hadoop nodes when running MRUniNovo. In the largest-scale experiment with a 12 GB dataset, MRUniNovo requires only 271 minutes on a 6-machine Hadoop cluster, which was shown devastating performance in terms of acceleration.

With the remarkable and rapid increase in spectrum data, the efficiency of *de novo* peptide sequencing of mass spectrometry has become a critical concern. Using distributed computing frameworks is a promising way to improve the performance of *de novo* peptide sequencing. MRUniNovo is a tool for parallel *de novo* peptide sequencing. The experimental results demonstrated that MRUniNovo could significantly reduce the computational time of *de novo* peptide sequencing. The ultimate aim of MRUninovo is to develop the pattern that can process the large-scale MS/MS data using HPC cluster in the future.

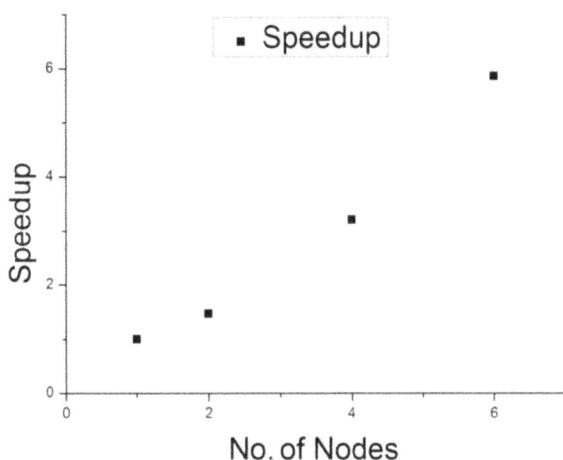

Fig. 8.7. Performance of MRUniNovo.

## 8.4. Conclusion

HPC has been playing a significant role in recent advances in bioinformatics, especially in MS/MS data analysis. In peptide identification,
due to benefiting from the advances in the mass spectrometry technology, the amount of MS/MS data has grown constantly and exponentially.
Thus, the peptide identification methods suffer from poor performance in
processing MS/MS data with low signal-to-noise ratio, posttranslational
modifications, and sequence variations. Nevertheless, the relatively slow
speed of MS/MS data analysis tool of today has lessened the significance of assisting peptide identification. In such cases, the data analysis
by HPC is a great way to improve efficiency. Since the HPC represen-
tatives are usually better suited for large-scale peptide identification than
most of the normal computer, the HPC can improve efficiency of peptide
identification.

Another way to improve efficiency of peptide identification is to use
deep learning and MS/MS data clustering. By using this combination of
deep learning and MS/MS data clustering, the MS/MS spectra representative's signal-to-noise ratio is improved by averaging the intensities of
related peaks in the cluster members, and its mass accuracy is improved by

averaging the masses of these peaks. Clustering has been extensively studied in machine learning and deep learning, include in terms of distance functions [47], feature selection [48–50], cluster validation [51], and grouping methods [52]. In [52, 53], author perform joint dimensionality reduction and clustering by first clustering the data with k-means and then projecting the data into a lower dimensions where the inter-cluster variance is maximized. Aggarwal and Reddy have [54] provided a comprehensive literature of deep learning of data clustering. MS/MS clustering replaces the huge amounts of raw MS/MS spectra by a much smaller number of representative MS/MS spectra. The main effect of spectrum clustering by deep learning is a tremendous reduction in data size, and the implication of this reduction on storage requirements, computing power needs, and analysis speed.

For bioinformatics, investigating proper methods to store, manage, and analyze MS/MS data, instead of human-processed features, and learn suitable features from those MS/MS spectra data is seen as a future challenge. In this respect, HPC has also won great success. In this work, we showed an extensive review of bioinformatics research applying HPC. We propose an efficient tool named MRUniNovo for parallel *de novo* peptide sequencing with the support of the MapReduce framework. Our experimental results demonstrate that MRUniNovo can significantly increase the efficiency of *de novo* peptide sequencing by HPC computation without sacrificing the computation accuracy. We further discussed shortcomings of the methods and trends of future research. Even with a perfect result in theory, HPC is still many defects and challenges, including interpretation of results, load imbalanced, and task scheduling. Furthermore, the improve accuracy through deep learning need further study and exploration to fully exploit the capabilities of deep learning. Thus, we believe that this work will serve as a starting point to explore the HPC and provide valuable insight to advance bioinformatics in future research.

## References

1. Manyika, J., Chui, M., Brown, B. *et al.* (2011). Big data: The next frontier for innovation, competition, and productivity. McKinsey Global Institute.
2. Weber, G. M., Mandl, K. D. and Kohane, I. S. (2014). Finding the missing link for big biomedical data. *The Journal of the American Medical Association* 311(24), pp. 2479–2480.

3.  Ye, J. and Liu, J. (2012). Sparse methods for biomedical data. *Journal of ACM Sigkdd Explorations Newsletter* 14(1), pp. 4–15.

4.  Ronneberger, O., Fischer, P. and Brox, T. (2015). U-net: Convolutional networks for biomedical image segmentation. In *International Conference on Medical Image Computing and Computer-assisted Intervention*, pp. 234–241. Springer, Cham.

5.  Schindelin, J., Arganda-Carreras, I., Frise, E., Kaynig, V., Longair, M., Pietzsch, T. and Tinevez, J. Y. (2012). Fiji: An open-source platform for biological-image analysis. *Nature Methods* 9(7), p. 676.

6.  Schindelin, J., Rueden, C. T., Hiner, M. C. and Eliceiri, K. W. (2015). The Image J. ecosystem: An open platform for biomedical image analysis. *Journal of Molecular Reproduction and Development* 82(7–8), pp. 518–529.

7.  Shameer, K., Badgeley, M. A., Miotto, R., Glicksberg, B. S., Morgan, J. W. and Dudley, J. T. (2016). Translational bioinformatics in the era of real-time biomedical, health care and wellness data streams. *Journal of Briefings in Bioinformatics*, p. 118.

8.  Agarwal, R. and Dhar, V. (2014). Big data, data science, and analytics: The opportunity and challenge for IS research. *Information Systems Research* 25(3), pp. 443–448.

9.  McLafferty, F. W. (1981). Tandem mass spectrometry. *Journal of Science* 214(4518), pp. 280–287.

10. Link, A. J., Eng, J., Schieltz, D. M., Carmack, E., Mize, G. J., Morris, D. R. and Yates, J. R. (1999). Direct analysis of protein complexes using mass spectrometry. *Journal of Nature Biotechnology* 17(7), p. 676.

11. Nesvizhskii, A. I., Keller, A., Kolker, E. and Aebersold, R. (2003). A statistical model for identifying proteins by tandem mass spectrometry. *Journal of Analytical Chemistry*, 75(17), pp. 4646–4658.

12. Hunt, D. F., Yates, J. R., Shabanowitz, J., Winston, S. and Hauer, C. R. (1986). Protein sequencing by tandem mass spectrometry. *Journal of Proceedings of the National Academy of Sciences* 83(17), pp. 6233–6237.

13. Li, C., Chen, T., He, Q., Zhu, Y. and Li, K. (2016). MRUniNovo: An efficient tool for *de novo* peptide sequencing utilizing the hadoop distributed computing framework. *Journal of Bioinformatics* 33(6), pp. 944–946.

14. Sanbonmatsu, K. Y. and Tung, C. S. (2007). High performance computing in biology: Multimillion atom simulations of nanoscale systems. *Journal of Structural Biology* 157(3), pp. 470–480.

15. Ciresan, D. C., Meier, U., Masci, J., Maria Gambardella, L. and Schmidhuber, J. (2011, July). Flexible, high performance convolutional neural networks for image classification. In *IJCAI Proceedings-International Joint Conference on Artificial Intelligence*, 22(1), p. 1237.

16. Feng, L. and Datta, A. (2011). A generic mutual information based algorithm for spatial registration of multi-modal medical images. *Journal of Medical Imaging and Health Informatics* 1(2), pp. 131–138.

17. Kalra, P. K. and Kumar, N. (2014). Poisson's equation based image registration: An application for matching 2D mammograms. *Journal of Medical Imaging and Health Informatics* 4(1), pp. 49–57.

18. Zhu, X., Li, K., Salah, A., Shi, L. and Li, K. (2015). Parallel implementation of MAFFT on CUDA-enabled graphics hardware. *IEEE/ACM Transactions on Computational Biology and Bioinformatics* 12(1), pp. 205–218.

19. Zhu, X., Li, K. and Salah, A. (2013). A data parallel strategy for aligning multiple biological sequences on multi-core computers. *Computers in Biology and Medicine* 43(4), pp. 350–361.

20. Dorband, J., Palencia, J. and Ranawake, U. (2003). Commodity computing clusters at goddard space flight center. *Journal of Space Communication* 1(3), pp. 113–123.

21. Brightwell, R., Fisk, L. A., Greenberg, D. S., Hudson, T., Levenhagen, M., MacCabe, A. B. and Riesen, R. (2000). Massively parallel computing using commodity components. *Parallel Computing* 26(2–3), pp. 243–266.

22. El-Ghazawi, T. A., Kaewpijit, S. and Le Moigne, J. (2001, October). Parallel and Adaptive Reduction of Hyperspectral Data to Intrinsic Dimensionality. In *Cluster* (p. 102).

23. Plaza, A., Valencia, D., Plaza, J. and Martinez, P. (2006). Commodity cluster-based parallel processing of hyperspectral imagery. *Journal of Parallel and Distributed Computing* 66(3), pp. 345–358.

24. Plaza, A., Plaza, J. and Valencia, D. (2007). Impact of platform heterogeneity on the design of parallel algorithms for morphological processing of high-dimensional image data. *The Journal of Supercomputing* 40(1), pp. 81–107.

25. Inoue, T., Yamatani, K., Itoh, K. and Ichioka, Y. (1993). Vector quantization for hyperspectral images using neural network. *Optics & Laser Technology* 25(3), p. 202.

26. Plaza, A., Valencia, D. and Plaza, J. (2008). An experimental comparison of parallel algorithms for hyperspectral analysis using heterogeneous and homogeneous networks of workstations. *Parallel Computing* 34(2), pp. 92–114.

27. Ekanayake, J. and Fox, G. (2009, October). High performance parallel computing with clouds and cloud technologies. In *International Conference on Cloud Computing* (pp. 20–38), Springer, Berlin, Heidelberg.

28. Tehranian, S., Zhao, Y., Harvey, T., Swaroop, A. and Mckenzie, K. (2006). A robust framework for real-time distributed processing of satellite data. *Journal of Parallel and Distributed Computing* 66(3), pp. 403–418.

29. Gnerre, S., MacCallum, I., Przybylski, D., Ribeiro, F. J., Burton, J. N., Walker, B. J. and Berlin, A. M. (2011). High-quality draft assemblies of mammalian genomes from massively parallel sequence data. *Proceedings of the National Academy of Sciences* 108(4), pp. 1513–1518.

30. Navarro, J., Vera, G., Ramos-Onsins, S. and Hernández, P. (2016, August). Improving Bioinformatics Analysis of Large Sequence Datasets Parallelizing Tools for Population Genomics. In *European Conference on Parallel Processing* (pp. 457–467), Springer, Cham.

31. Botterill, J. J., Fournier, N. M., Guskjolen, A. J., Lussier, A. L., Marks, W. N. and Kalynchuk, L. E. (2014). Amygdala kindling disrupts trace and delay fear conditioning with parallel changes in Fos protein expression throughout the limbic brain. *Neuroscience* 265, pp. 158–171.

32. Macosko, E. Z., Basu, A., Satija, R., Nemesh, J., Shekhar, K., Goldman, M. and Trombetta, J. J. (2015). Highly parallel genome-wide expression profiling of individual cells using nanoliter droplets. *Cell* 161(5), pp. 1202–1214.

33. Marcotte, E. M. (2007). How do shotgun proteomics algorithms identify proteins? *Nature Biotechnology* 25(7), p. 755.

34. Yang, H., Chi, H., Zhou, W. J., Zeng, W. F., He, K., Liu, C. and He, S. M. (2017). Open-pNovo: de novo peptide sequencing with thousands of protein modifications. *Journal of Proteome Research* 16(2), pp. 645–654.

35. Sadygov, R. G., Cociorva, D. and Yates III, J. R. (2004). Large-scale database searching using tandem mass spectra: looking up the answer in the back of the book. *Nature Methods* 1(3), p. 195.

36. Kapp, E. and Schütz, F. (2007). Overview of tandem mass spectrometry (MS/MS) database search algorithms. *Current Protocols in Protein Science*, pp. 25–32.

37. Eng, J. K., McCormack, A. L. and Yates, J. R. (1994). An approach to correlate tandem mass spectral data of peptides with amino acid sequences in a protein database. *Journal of the American Society for Mass Spectrometry* 5(11), pp. 976–989.

38. Sadeh, N. M., Hildum, D. W., Kjenstad, D. and Tseng, A. (1999). Mascot: An agent-based architecture for coordinated mixed-initiative supply chain planning and scheduling. In Workshop on Agent-Based Decision Support in Managing the Internet-Enabled Supply-Chain, at Agents' 99.

39. Bjornson, R. D., Carriero, N. J., Colangelo, C., Shifman, M., Cheung, K. H., Miller, P. L. and Williams, K. (2007). X!! Tandem, an improved method for running X! tandem in parallel on collections of commodity computers. *The Journal of Proteome Research* 7(1), pp. 293–299.

40. Allmer, J. (2011). Algorithms for the *de novo* sequencing of peptides from tandem mass spectra. *Expert Review of Proteomics* 8(5), pp. 645–657.

41. Ma, B., Zhang, K., Hendrie, C., Liang, C., Li, M., Doherty-Kirby, A. and Lajoie, G. (2003). PEAKS: powerful software for peptide de novo sequencing by tandem mass spectrometry. *Rapid Communications in Mass Spectrometry* 17(20), pp. 2337–2342.

42. Jeong, K., Kim, S. and Pevzner, P. A. (2013). UniNovo: A universal tool for *de novo* peptide sequencing. *Bioinformatics* 29(16), pp. 1953–1962.

43. Pratt, B., Howbert, J. J., Tasman, N. I. and Nilsson, E. J. (2011). MR-tandem: Parallel X! tandem using hadoop MapReduce on amazon Web services. *Bioinformatics* 28(1), pp. 136–137.

44. Craig, R. and Beavis, R. C. (2004). TANDEM: Matching proteins with tandem mass spectra. *Bioinformatics* 20(9), pp. 1466–1467.

45. Bjornson, R. D., Carriero, N. J., Colangelo, C., Shifman, M., Cheung, K. H., Miller, P. L. and Williams, K. (2007). X!! Tandem, an improved method for running X! tandem in parallel on collections of commodity computers. *The Journal of Proteome Research* 7(1), pp. 293–299.

46. Jeong, K., Kim, S. and Pevzner, P. A. (2013). UniNovo: A universal tool for *de novo* peptide sequencing. *Bioinformatics* 29(16), pp. 1953–1962.

47. Xiang, S., Nie, F. and Zhang, C. (2008). Learning a Mahalanobis distance metric for data clustering and classification. *Journal of Pattern Recognition* 41(12), pp. 3600–3612.

48. Liu, H. and Yu, L. (2005). Toward integrating feature selection algorithms for classification and clustering. *IEEE Transactions on Knowledge and Data Engineering* 17(4), pp. 491–502.

49. Boutsidis, C., Drineas, P. and Mahoney, M. W. (2009). Unsupervised feature selection for the $k$-means clustering problem. In *Advances in Neural Information Processing Systems*, pp. 153–161.

50.  Liu, H. and Yu, L. (2005). Toward integrating feature selection algorithms for classification and clustering. *IEEE Transactions on Knowledge and Data Engineering* 17(4), pp. 491–502.

51.  Halkidi, M., Batistakis, Y. and Vazirgiannis, M. (2001). On clustering validation techniques. *Journal of Intelligent Information Systems* 17(2–3), pp. 107–145.

52.  Von Luxburg, U. (2007). A tutorial on spectral clustering. *Journal of Statistics and Computing* 17(4), pp. 395–416.

53.  Ye, J., Zhao, Z. and Wu, M. (2008). Discriminative k-means for clustering. In *Advances in Neural Information Processing Systems*, pp. 1649–1656.

54.  De la Torre, F. and Kanade, T. (2006, June). Discriminative cluster analysis. In *Proceedings of the 23rd International Conference on Machine Learning*, pp. 241–248, ACM.

Chapter 9

# Analysis of Boolean Networks and Boolean Models of Metabolic Networks

Tatsuya Akutsu

*Bioinformatics Center, Institute for Chemical Research,*
*Kyoto University, Kyoto 611-0011, Japan*

## 9.1. Introduction

Analysis of biological networks is an important topic in bioinformatics and systems biology. For that purpose, various kinds of mathematical models have been proposed and applied. Among them, the *Boolean network* (BN) has been extensively studied since 1960s [1]. In a BN, each node corresponds to a gene and takes either 0 or 1 at each time step, where 1 (resp., 0) means that the corresponding gene is expressed (resp., not expressed). The states of nodes change synchronously according to regulation rules given as Boolean functions. Although the BN has been mainly used as a model of genetic networks, it has also been used for modeling various biological systems, including metabolic networks and neural networks. BNs exhibit complex behavior and can explain various phenomena of biological systems. Therefore, many studies have been done on the development of computational methods for analysis of BN models [2] and on BN-based modeling of various biological systems [3, 4].

In this chapter, we focus on two problems on BNs: *discrimination of attractors* [5] and *reaction cut in metabolic networks* [6]. The first one is a problem of finding a minimum set of nodes so that given attractors are discriminated by looking at time-series expression (0-1) data of these nodes. The problem might be useful for finding gene markers to discriminate certain types of cells because attractors are often interpreted as cell types. The second one is a problem of finding a set of reaction nodes (corresponding to chemical reactions) so that the specified (non-preferable) compounds are not producible. This problem and its variants might be useful to increase or decrease the efficiency of producing useful or harmful products by knockout of genes.

The organization of this chapter is as follows. First, we introduce the BN and related concepts (e.g. attractors). Next, we review the problem and method for discrimination of attractors and present its application to realistic models of biological networks. Then, we review the problems, methods, extensions, and some case studies of reaction cut in metabolic networks. Finally, we conclude with discussions on future research directions.

## 9.2.  Boolean Networks

A BN consists of a set of nodes and a set of Boolean functions, where one function is assigned to each node. Formally, a BN $N(V, F)$ consists of a set $V = \{x_1, \ldots, x_n\}$ of nodes and a list $F = (f_1, \ldots, f_n)$ of *Boolean functions*, where a Boolean function $f_i(x_{i_1}, \ldots, x_{i_{k_i}})$ with inputs from specified nodes $x_{i_1}, \ldots, x_{i_{k_i}}$ is assigned to each node $x_i$. For each $x_1$, the set of input nodes $x_{i_1}, \ldots, x_{i_k}$ to $x_i$ is denoted by $IN(x_i)$. Each node takes either 0 or 1 at each discrete time $t$, and the state of node $x_i$ at time $t$ is denoted by $x_i(t)$, where 1 (resp., 0) means that the corresponding gene is expressed (resp., not expressed). The state of node $x_i$ at time $t + 1$ is determined by

$$x_i(t + 1) = f_i(x_{i_1}(t), \ldots, x_{i_{k_i}}(t)),$$

and the state of the whole network at time step $t$ is represented by an $n$-dimensional 0-1 vector $\mathbf{x}(t) = [x_1(t), \ldots, x_n(t)]$. We also write

$$x_i(t + 1) = f_i(\mathbf{x}(t))$$

to denote the regulation rule for $x_i$ and

$$\mathbf{x}(t+1) = \mathbf{f}(\mathbf{x}(t))$$

to denote the regulation rule for the whole BN. The network structure of $N(V, F)$ is represented by a directed graph $G(V, E)$ defined by

$$E = \{(x_{i_j}, x_i) | x_{i_j} \in IN(x_i)\}.$$

The dynamics of a BN is well-described by a *state transition diagram*, in which a vertex and a directed edge correspond to a (global) state of the BN and a state transition, respectively. Note that in this chapter, we use nodes and vertices for BNs and state transition diagrams, respectively. Consider a BN $N(V, F)$ defined by

$$x_1(t+1) = x_3(t)$$
$$x_2(t+1) = x_1(t) \wedge \overline{x_3(t)}$$
$$x_3(t+1) = x_1(t) \wedge \overline{x_2(t)}$$

where $x \wedge y$ and $\overline{x}$ denote AND of $x$ and $y$ and NOT of $x$, respectively. Then, $G(V, E)$ and its state transition diagram are given as in Figs. 9.1(a) and (b), respectively.

Starting from any initial state, a BN will eventually reach a cyclic sequence of states. This sequence of states is called an *attractor*. An attractor is often interpreted as a type of a cell: different attractors correspond to different cell types. An attractor consisting of only one state (i.e. $\mathbf{x} = \mathbf{f}(\mathbf{x})$) is called a *singleton attractor*. Otherwise, it is called a *periodic attractor*.

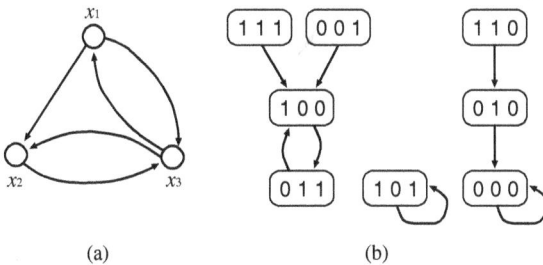

Fig. 9.1. Example of BN. (a) $G(V, E)$. (b) State transition diagram.

A periodic attractor consisting of $p$ states is called a $p$-periodic attractor. For example, the BN given in Fig. 9.1 has two singleton attractors

$$\langle [0, 0, 0] \rangle, \quad \langle [1, 0, 1] \rangle$$

and one 2-periodic attractor

$$\langle [0, 1, 1], [1, 0, 0] \rangle.$$

It is to be noted that a $p$-periodic attractor can have $p$ representations. For example, the 2-periodic attractor in Fig. 9.1 has two representations:

$$\langle [0, 1, 1], [1, 0, 0] \rangle, \quad \langle [1, 0, 0], [0, 1, 1] \rangle.$$

Since attractors play key roles in analysis of BNs, extensive studies have been both on detection/enumeration of attractors and on the distribution of attractors [2].

## 9.3. Discrimination of Attractors

In this section, we consider the problem of determining the minimum set of nodes by which all given singleton and periodic attractors can be discriminated. In this problem, we assume that a set of attractors is given as an input and a BN is not explicitly given. This assumption is reasonable when we need to find marker genes that are required to discriminate given types of cells using gene expression data, because a precise gene regulatory network structure is usually unknown.

### 9.3.1. *Discrimination of Singleton Attractors*

We begin with the case of discrimination of singleton attractors. This problem is formally defined as follows. Let $B$ be an $m \times n$ binary matrix, where a column and a row correspond to a node and a singleton attractor, respectively. $B[i, j]$ denotes the element at $i$th row and $j$th column, $B[i, -]$ denotes the $i$th row, and $B[-, j]$ denotes the $j$th column. Let $J = \{j_1, \ldots, j_k\}$ be a set of column indices. Then, $B_J$ denotes the submatrix of $B$ consisting of the $j_1, j_2, \ldots, j_k$th columns. The discrimination problem of singleton attractors is defined as follows.

**Definition 9.1. [Discrimination of Singleton Attractors]**
Instance: An $m \times n$ binary matrix where each column corresponds to a node
and each row corresponds to a singleton attractor.

Problem: Find a minimum cardinality set $J$ of columns such that
$B_J[i_1, -] \neq B_J[i_2, -]$ holds for all $i_1, i_2$ with $1 \leq i_1 < i_2 \leq m$.

For example, consider a $6 \times 7$ matrix $B$ defined by

$$B = \begin{pmatrix} 0 & 1 & 1 & 1 & 0 & 1 & 1 \\ 1 & 1 & 1 & 1 & 0 & 0 & 0 \\ 1 & 0 & 1 & 1 & 0 & 1 & 0 \\ 0 & 0 & 1 & 0 & 1 & 1 & 1 \\ 1 & 0 & 1 & 0 & 1 & 1 & 1 \\ 0 & 1 & 1 & 1 & 0 & 0 & 0 \end{pmatrix}.$$

Note that each column corresponds to a gene and each row corresponds
to an attractor (i.e. cell type). Then, $B[1, -] = [0, 1, 1, 1, 0, 1, 1]$ and
$B[-, 2] = [1, 1, 0, 0, 0, 1]^\top$, where $A^\top$ denotes the transposed matrix of
$A$. $J = \{1, 5, 6\}$ is a solution of Discrimination of Singleton Attractors for
$B$ because rows are mutually different in

$$B_{\{1,5,6\}} = \begin{pmatrix} 0 & 0 & 1 \\ 1 & 0 & 0 \\ 1 & 0 & 1 \\ 0 & 1 & 1 \\ 1 & 1 & 1 \\ 0 & 0 & 0 \end{pmatrix}$$

and there is no such index set of size 2.

This problem is a kind of feature selection problem, which has been
extensively studied in machine learning. It is known that this problem is
NP-hard whereas it can be solved in $O((m/1.146)^m poly(m, n))$ time using
a dynamic programming algorithm. Readers interested in details of the
algorithm are referred to [5].

### 9.3.2. *Discrimination of Singleton and Periodic Attractors*

Next, we consider all attractors: singleton and periodic attractors. We need some definitions before formally defining the problem. For a set $U \subseteq V$ and a 0-1 vector $\mathbf{x}$ for $V = \{x_1, \ldots, x_n\}$, $\mathbf{x}_U$ denotes the $|U|$-dimensional vector consisting of elements of $\mathbf{x}$ that correspond to $U$. For example, for $\mathbf{x} = [1, 1, 0, 0, 1, 0]$, $\mathbf{x}_U = [1, 1, 0]$ for $U = \{x_1, x_2, x_3\}$, and $\mathbf{x}_U = [1, 0, 1, 0]$ for $U = \{x_2, x_4, x_5, x_6\}$.

Let

$$A_h = \langle \mathbf{x}(0), \mathbf{x}(1), \ldots, \mathbf{x}(p(A_h) - 1) \rangle$$

be an attractor of period $p(A_h)$ (i.e. $\mathbf{x}(0) = \mathbf{x}(p(A_h))$) on an underlying BN. Let $Ser(A_h, U, t)$ be the infinite sequence of vectors defined by

$$Ser(A_h, U, t) = \langle \mathbf{x}_U(t), \mathbf{x}_U(t+1), \mathbf{x}_U(t+2), \ldots \rangle.$$

Two periodic attractors $A_h$ and $A_k$ are *identical* if and only if $Ser(A_h, V, 0) = Ser(A_k, V, t)$ holds for some $t \geq 0$.

**Definition 9.2.   [Discrimination of Attractors]**
Instance: A set of attractors $\mathcal{A} = \{A_1, A_2, \ldots, A_m\}$, where each $A_h$ consists of $p(A_h)$ states represented as a sequence of $p(A_h)$ binary vectors, $p(A_h)$ denotes the period of $A_h$, $n > 0$ denotes the number of nodes, and $m > 1$.

Problem: Find a minimum cardinality set $U$ of nodes such that $Ser(A_h, U, 0) \neq Ser(A_k, U, t)$ for any $t \geq 0$ when $h \neq k$.

**Example 9.1.** Let $\mathcal{A} = \{A_1, A_2, A_3, A_4, A_5\}$ be a set of attractors, where

$$A_1 = \langle [1, 1, 1, 0, 0] \rangle$$
$$A_2 = \langle [0, 1, 0, 0, 1] \rangle$$
$$A_3 = \langle [0, 1, 1, 0, 1] \rangle$$
$$A_4 = \langle [1, 0, 0, 0, 0], [0, 1, 1, 0, 1], [1, 1, 0, 1, 0], [0, 0, 1, 1, 1] \rangle$$
$$A_5 = \langle [0, 1, 1, 0, 0], [1, 1, 0, 1, 1], [0, 0, 1, 1, 0], [1, 0, 0, 0, 1] \rangle.$$

Then, $\{x_1, x_3, x_5\}$ is a solution of Discrimination of Attractors (see also Fig. 9.2) since $A_1$, $A_2$, and $A_3$ can be distinguished by looking at

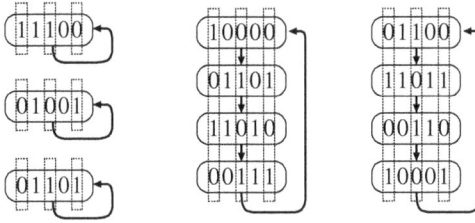

Fig. 9.2. Example of Discrimination of Attractors. In this example, there are three singleton attractors and two periodic attractors, all of which can be discriminated by looking at time series data of three genes shown by dotted boxes.

$x_1$ and $x_3$, and $Ser(A_4, \{x_1, x_3, x_5\}, 0) \neq Ser(A_5, \{x_1, x_3, x_5\}, t)$ holds for any $t$. Note that $\{x_1, x_2, x_3\}$ is not a solution since

$$Ser(A_4, \{x_1, x_2, x_3\}, 0) = Ser(A_5, \{x_1, x_2, x_3\}, 3)$$
$$= \langle [1, 0, 0], [0, 1, 1], [1, 1, 0], [0, 0, 1], \dots, \rangle$$

holds.

This example suggests the difficulty of selecting a set of nodes for discriminating even a pair of periodic attractors. However, the following lemma states that we can always select two nodes that discriminate two given periodic attractors, where it is proved using Chinese Remainder Theorem.

**Lemma 9.1.** *For any two distinct (i.e. non-identical) attractors* $A_h = \langle \mathbf{x}(0), \dots, \mathbf{x}(p_h - 1) \rangle$ *and* $A_k = \langle \mathbf{w}(0), \dots, \mathbf{w}(p_k - 1)) \rangle$, *there exists* $U \subseteq V$ *of* $|U| = 2$ *for which* $Ser(A_h, U, 0) \neq Ser(A_k, U, t)$ *holds for any* $t \geq 0$.

It is easy to see that one node is enough to discriminate a singleton attractor and a singleton or periodic attractor. Therefore, the following number gives an upper bound of the number of nodes to discriminate all $m$ attractors:

$$2 \cdot \binom{m}{2} = m(m - 1).$$

This observation leads to a simple exhaustive-type algorithm for Discrimination of Attractors. Recall that $\mathcal{A} = \{A_1, \dots, A_m\}$ is a given set of attractors. Let $p$ be the longest period among $A_1, \dots, A_m$. We modify

$Ser(A_h, U, t)$ so that its length becomes finite. Let $Ser(A_h, U, t, p)$ be the sequence of vectors of length $p$ defined by

$$Ser(A_h, U, t, p) = \langle \mathbf{x}_U(t), \mathbf{x}_U(t+1), \mathbf{x}_U(t+2), \ldots, \mathbf{x}_U(t+p-1) \rangle.$$

The following is a pseudocode of the resulting algorithm for Discrimination of Attractors, which returns a minimum set of discrimination nodes $U$:

> **Procedure** Discrimination of Attractors($\mathcal{A}$)
> **for** $s = 1$ **to** $m(m-1)$ **do**
>   **for all** $U \subseteq V$ such that $|U| = s$ **do**
>     $flag \leftarrow 1$;
>     **for all** $(A_i, A_j)$ such that $i < j$ **do**
>       **if** $Ser(A_i, U, 0, p) = Ser(A_j, U, t, p)$ holds for some
>       $t \in \{0, \ldots, p-1\}$
>       **then** $flag \leftarrow 0$; **break**;
>     **if** $flag = 1$ **return** $U$.

By using this algorithm, we can show the following theorem, where the proof is given in [5].

**Theorem 9.1.** *Discrimination of Attractors can be solved in $O(n^{|U_{\min}|}$ $poly(m, n, p))$ time, where $U_{\min}$ is an optimal solution and $p$ is the longest period of input attractors. Furthermore, $|U_{\min}| \leq m(m-1)$ holds.*

### 9.3.3. Case Study on Discrimination of Attractors

The developed algorithms were applied to BN models of the following four biological processes:

(1) Expression patterns of the segment polarity genes in *Drosophila Melanogaster* [7]. The corresponding network includes 60 genes (15 genes each for four different cells) and 10 singleton attractors.
(2) Control of the mammalian cell cycle [8]. The corresponding network includes 10 genes, one singleton attractor, and one periodic attractor of length 7.

(3) Logical model analyzing T-cell activation [9]. The corresponding network includes 40 genes, 8 singleton attractors, and one periodic attractor of length 6.
(4) Boolean model of IGVH mutational status in chronic lymphocytic leukemia [10]. The corresponding network includes 90 genes and 6 periodic attractors all with period 4.

Then, the following genes were selected as minimum sets of discrimination nodes:

(1) $wg_1$, $wg_2$, $PTC_1$, $SMO_2$, $en_2$
(2) CycD
(3) CD45, CD8, TCRbind
(4) AEBP1, CCND2, INPP5D

It is seen that a small number of genes are enough to discriminate the above attractors. Furthermore, it is found that many of these genes play important roles in the four biological processes, where details are given in [5].

## 9.4. Analysis of Metabolic Networks

In this section, we review a Boolean model of metabolic networks and linear integer programming-based (ILP-based) methods for the model.

### 9.4.1. *Boolean Model of Metabolic Networks*

Chemical compounds and chemical reactions play important roles in cells. Relations on chemical compounds and chemical reactions are represented as metabolic networks. Although a number of studies have been done on analyses of metabolic networks under the flux balance analysis (FBA) framework [11, 12], studies using Boolean models have also been done [13–15]. In most of Boolean models, a metabolic network contains two kinds of nodes: compound nodes and reaction nodes. As in BNs, each node takes either 0 or 1, where 0 and 1 mean that the corresponding compounds/reactions are active and inactive, respectively. Furthermore, reaction nodes and compound nodes are modeled as AND nodes and OR nodes, respectively, because all of substrates must be present in order to activate a chemical reaction whereas some amount of specific chemical compound can be generated if one of the chemical reactions generating the compound is active. Accordingly, a Boolean model (BN model) of metabolic network is defined as follows.

Let $V_c = \{v_{c_1}, \ldots, v_{c_m}\}$ and $V_r = \{v_{r_1}, \ldots, v_{r_n}\}$ be a set of *compound nodes* and a set of *reaction nodes*, respectively, where $V_c \cap V_r = \emptyset$ and we let $V = V_c \cup V_r$. In order to apply the model to a kind of knockout problems, we consider two kinds of special nodes. Let $V_s \subseteq V_c$ and $V_t \subseteq V_c$ be a set of *source nodes* and a set of *target nodes*, respectively, where $V_s \cap V_t = \emptyset$. Then, a metabolic network is defined as a directed graph $G(V, E)$ satisfying the following conditions:

- For each edge $(u, v) \in E$, either $(u \in V_c) \wedge (v \in V_r)$ or $(u \in V_r) \wedge (v \in V_c)$ holds.
- Each node $v \in V_s$ does not have an incoming edge.
- Each node $v \notin V_s$ has at least one incoming edge.

In this section, we mainly consider the problem of finding a minimum set of reactions knockout of which prevents production of a specified subset of the target compounds. This problem is referred to as the *reaction cut problem* in this section. Let $V_d \subseteq V_r$ be a set of reaction nodes that are to be knocked out. In the reaction cut problem, we only consider stable states of metabolic networks and thus we do not consider time steps. Accordingly, we write $v = 1$ (resp., $v = 0$) if 1 (resp., 0) is assigned to a node $v \in V$. Let $A$ be an assignment of 0-1 values to nodes in $V$. $A$ is called a *valid assignment* if the following conditions are satisfied.

- For each $v \in V_s$, $v = 1$.
- For each $v \in V_c - V_s$, $v = 1$ iff there is $u$ such that $(u.v) \in E$ and $u = 1$.
- For each $v \in V_r$, $v = 1$ iff $v \notin V_d$ holds and $u = 1$ holds for all $a$ such that $(a, v) \in E$.

The first, second, and third conditions state respectively that the source compounds are always available, compound nodes correspond to OR nodes, and reaction nodes correspond to AND nodes where the state of a node $v$ is forced to be 0 if $v \in V_d$.

## Definition 9.3. [Reaction Cut]

Instance: A BN-model of a metabolic network $G(V, E)$ with $V = V_c \cup V_r$ and $V_s \subseteq V_c$, and a set of target compounds $V_t \subseteq V_c$ such that $V_t \cap V_s = \emptyset$.

Problem: Find a minimum cardinality set of reactions $V_d \subseteq V_r$ knockout of which prevents production of any compound of $V_t$ (i.e. $v = 0$ for all $v \in V_t$) under any valid assignment.

$V_d$ satisfying the above condition (without minimality) is called a *reaction cut*. It should be noted that since $V_d = V_r$ is always a reaction cut, Reaction Cut always has a solution.

For example, consider a BN model of a metabolic network in Fig. 9.3 (a), where $V_c = \{v_1, v_2, v_3, v_4, v_5, v_6, v_7, v_8, v_9\}$, $V_r = \{v_a, v_b, v_c, v_d\}$, $V_s = \{v_1, v_2, v_3, v_4\}$, and $V_t = \{v_9\}$. If we select $V_d = \{v_a\}$ in BN (A), we have $v_1 = 1, v_2 = 1, v_3 = 1, v_4 = 1, v_a = 0, v_b = 1, v_5 = 0, v_6 = 0, v_7 = 1, v_8 = 1, v_c = 0, v_d = 0, v_9 = 0$. Therefore, $\{v_a\}$ is a reaction cut of BN (A). Furthermore, it is a minimum reaction cut because $v_9 = 1$ holds if all reactions are active (i.e. $V_d = \emptyset$). Similarly, we can see that $\{v_b\}$ is also a minimum reaction cut.

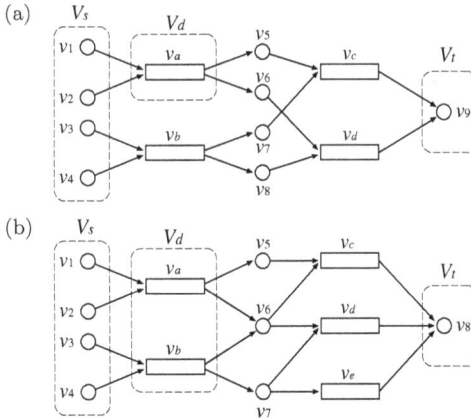

Fig. 9.3. Examples of BN models of metabolic networks. Circles and rectangles represent compound nodes (i.e. OR nodes) and reaction nodes (i.e. AND nodes), respectively. $V_d = \{v_a\}$ is a minimum reaction cut in (a), whereas $V_d = \{v_a, v_b\}$ is a minimum reaction cut in (b).

Next, consider a BN model in Fig. 9.3(b). If we select $V_d = \{v_a\}$ in BN (B), we have $v_1 = 1, v_2 = 1, v_3 = 1, v_4 = 1, v_a = 0, v_b = 1, v_5 = 0,$ $v_6 = 1, v_7 = 1, v_c = 0, v_d = 1, v_e = 1, v_8 = 1$. Therefore, $\{v_a\}$ is not a reaction cut of BN (B). Similarly, we can see that $\{v_b\}$ is not a reaction cut. If we select $V_d = \{v_a, v_b\}$, we have $v_1 = 1, v_2 = 1, v_3 = 1, v_4 = 1, v_a = 0,$ $v_b = 0, v_5 = 0, v_6 = 0, v_7 = 0, v_c = 0, v_d = 0, v_e = 0, v_8 = 0$. Therefore, $\{v_a, v_b\}$ is a reaction cut of BN (B). Since $v_8 = 1$ holds for any $V_d$ such that $|V_d| \leq 1$, $\{v_a, v_b\}$ is a minimum reaction cut of BN (B).

### 9.4.2. *ILP-based Method for Reaction Cut*

It is known that Reaction Cut is NP-hard [6], which suggests that it is not plausible that there exists a polynomial time algorithm for Reaction Cut. In order to cope with NP-hardness, various approaches can be considered. Use of *integer linear programming* (ILP) is one of the major approaches to practically solving NP-hard problems because several practically efficient ILP solvers are available and many NP-hard problems have simple ILP formulations.

Before presenting an ILP-based method for Reaction Cut, we briefly review ILP. ILP is based on *linear programming* (LP), where LP is a problem/framework for optimizing (minimizing or maximizing) a linear objective function subject to linear inequality and equality constraints. In the minimization version of LP, an instance is given as a linear program:

$$\text{minimize} \quad c_1 x_1 + \cdots + c_n x_n$$

$$\text{subject to} \quad a_{i,1} x_1 + \cdots + a_{i,n} x_n \leq b_i, \quad \text{for } i = 1, \ldots, m$$

$$x_i \geq 1, \quad \text{for } i = 1, \ldots, n$$

where $x_i$s are variables, and $a_{i,j}$s, $b_i$s, and $c_i$s are given coefficients (i.e. constants). Each $[x_1, \ldots, x_n]$ satisfying all constraints is called a *feasible solution*, and a feasible solution attaining the optimum objective value is called an *optimal solution*. ILP is defined in the same way except that all variables must take integer values. Accordingly, optimal solutions may differ between LP and ILP. If some specified variables must take integer values and the other variables can take real values, it is called *mixed integer linear programming* (MILP). It is known that LP is solvable in polynomial time whereas ILP (resp., MILP) is NP-hard.

For example, consider the following instance of LP:

$$\text{minimize} \quad x_1 + x_2$$

$$\text{subject to} \quad 2x_1 + x_2 \geq 2$$

$$x_1 + 2x_2 \geq 2.$$

Then, the objective function takes the minimum value of $4/3$ at $[x_1, x_2] = [2/3, 2/3]$. On the other hand, in ILP, the objective function takes the minimum value of 2 at $[x_1, x_2] = [2, 0]$, $[x_1, x_2] = [1, 1]$, or $[x_1, x_2] = [0, 2]$.

Here, we explain an ILP-based formulation of Reaction Cut. In this formulation, every variable $x$ takes either 0 or 1. First, we explain AND and OR functions can be represented using linear inequalities with integer constraints. Let $y = x_1 \wedge \cdots \wedge x_k$ (i.e. $y$ is AND of $x_1, \ldots, x_k$). It is not difficult to see that this constraint can be represented as

$$ky \leq x_1 + \cdots + x_k$$

$$y \geq x_1 + \cdots + x_k - (k - 1).$$

Analogously, let $y = x_1 \vee \cdots \vee x_k$ (i.e. $y$ is OR of $x_1, \ldots, x_k$). Then, this constraint can be represented as

$$ky \geq x_1 + \cdots + x_k$$

$$y \leq x_1 + \cdots + x_k.$$

By using the above formulations, we can represent constraints on compound nodes and reaction nodes in the ILP form. For each source node $v_s$, we put a constraint that $v_s = 1$. Analogously, for each target node $v_t$, we put a constraint that $v_t = 0$. Finally, the objective function is defined as

$$\text{minimize} \quad \sum_{i=1}^{m}(1 - x_{r_i})$$

which means that the number of reaction nodes to be knocked out must be minimized, where $x_{r_i}$ is a binary variable corresponding to a reaction node $r_i$, and $x_{r_i} = 0$ (i.e. $1 - x_{r_i} = 1$) means that the reaction $r_i$ is knocked out.

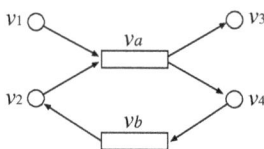

Fig. 9.4. Example of a BN model of a metabolic network with a cycle, where $V_c = \{v_1, v_2, v_3, v_4\}$, $V_s = \{v_1\}$, $V_t = \{v_3\}$, and $V_r = \{v_a, v_b\}$. There exist two valid assignments for this BN: $[v_1, v_2, v_3, v_4, v_a, v_b] = [1, 1, 1, 1, 1, 1]$ and $[v_1, v_2, v_3, v_4, v_a, v_b] = [1, 0, 0, 0, 0, 0]$, among which the former one is the maximal one.

Although the states of nodes are uniquely determined for given $V_d$ in acyclic networks, these are not necessarily so in cyclic networks. For example, consider a BN model given in Fig. 9.4. For this BN, there exist two valid assignments: $[v_1, v_2, v_3, v_4, v_a, v_b] = [1, 1, 1, 1, 1, 1]$ and $[v_1, v_2, v_3, v_4, v_a, v_b] = [1, 0, 0, 0, 0, 0]$. If we adopt the latter assignment, $V_d = \emptyset$ becomes a solution for Reaction Cut. However, the former assignment does not correspond to a solution. In order to avoid such an ambiguity, the concept of a *maximal valid assignment* was introduced [6]. An assignment $A$ is called *maximal* if $A[v] \geq A'[v]$ holds for all nodes $v \in (V_c \cup V_r)$ under any valid assignment $A'$, where $A[v]$ denotes the 0-1 value of node $v$ under $A$. It is shown in [6] that the maximal valid assignment always exists and is uniquely determined for every BN model of a metabolic network. For the BN model in Fig. 9.4, the maximal valid assignment is $[v_1, v_2, v_3, v_4, v_a, v_b] = [1, 1, 1, 1, 1, 1]$. In order to prevent production of compound $v_3$, it is enough to let $V_d = \{v_a\}$ or $V_d = \{v_b\}$, each of which gives the maximal valid assignment of $[v_1, v_2, v_3, v_4, v_a, v_b] = [1, 0, 0, 0, 0, 0]$. Computation of $V_d$ under the constraint of the maximal valid assignment can be done by modifying the ILP formulation with introducing time steps, where details are given in [6].

### 9.4.3. *Case Study on Metabolic Networks*

We applied one of the developed ILP-based methods for Reaction Cut to metabolic subnetworks of three organisms: *Homo sapiens* (hsa), *Saccharomyces cerevisiae* (sce), and *E. coli* (eco). For eco, eco00010.xml, eco00020.xml, eco00030.xml, eco00040.xml, eco00051.xml, eco00052.xml, eco00053.xml, eco00061.xml, eco00062.xml, eco00071.xml, eco00100.xml, eco00120.xml, eco00130.xml of KEGG database [16] were combined, and the resulting network consisted of 261 compound nodes and 404 reaction nodes. For sce, sce00010.xml, sce00020.xml, sce00030.xml, sce00040.xml, sce00051.xml, sce00052. xml, sce00061.xml, sce00062.xml, sce00071.xml, sce00072.xml, sce00 100.xml, sce00120.xml, sce00130.xml, of KEGG database were combined, and the resulting network consisted of 198 compound nodes and 286 reaction nodes. For hsa, hsa00010.xml, hsa00020.xml, hsa00030.xml, hsa00040.xml, hsa00051.xml, hsa00052.xml, hsa00053.xml, hsa00061. xml, hsa00062.xml, hsa00071.xml, hsa00072.xml, hsa00100.xml, hsa 00120.xml, hsa00130.xml of KEGG database were combined, and the resulting network consisted of 271 compound nodes and 410 reaction nodes. Pyruvate (C00022), Acetyl-CoA (C00024), Acetate(C00033), Oxaloacetate (C00036), and Phosphoenolpyruvate (C00074) were used as target compounds from a viewpoint of importance of amino acids.

The result is summarized in Table 9.1. The biological meanings of the selected reactions were analyzed in [6] and it was shown that important reactions for the synthesis of amino acids and the production of biogenetic energy were identified. It may look curious that the number of inactivated reactions of *E. coli* (prokaryotes) is larger than that of yeast and human (eukaryotes). However, this result is reasonable because eukaryotes do not have any metabolic pathways such as the ED pathway and the ascorbate metabolism.

The proposed ILP-based approach was further extended as follows:

- Minimum Reaction Insertion (MRI) [17]: find the minimum set of additional reactions from a reference metabolic network to a host metabolic network so that a target compound becomes producible in the revised

Table 9.1. Results on Reaction Cut for metabolic networks of three organisms.

|  | C00022 | C00024 | C00033 | C00036 | C00074 | All |
|---|---|---|---|---|---|---|
| *E.coli* (eco) | 3 | 3 | 2 | 3 | 2 | 6 |
| yeast (sce) | 1 | 2 | 2 | 2 | 2 | 3 |
| human (hsa) | 2 | 3 | 2 | 4 | 3 | 5 |

host metabolic network. The method was applied to metabolic networks of *E. coli* and reference networks from KEGG database.

- Minimal Knockout for Multiple Networks (MKMN) [18]: find the minimum set of reactions whose inhibition induces the target compounds to become non-producible in bad cells but producible in good cells. The method was applied to the metabolic networks data of *clostridium perfringens SM101* (corresponding to a bad organism) *bifidobacterium longum DJO10A* (corresponding to a good organism).

- Boolean Reaction Modification (BRM) [19]: minimize the total number of removed reactions from the host network and added reactions from the reference network so that the toxic compounds are not producible, but the necessary compounds are producible in the resulting host network. The method was applied to metabolic networks of *E. coli* and reference networks from KEGG database.

## 9.5.   Concluding Remarks

In this chapter, we reviewed two applications of the Boolean network (BN): discrimination of attractors [5] and reaction cut in metabolic networks [6]. The former one is a problem of finding a minimum set of nodes so that given attractors (corresponding to cell types) are discriminated by looking at time-series binary expression data of these nodes. The problem might be useful for finding gene markers to discriminate certain types of cells. The result of computational experiments suggests that a small number of genes are enough to discriminate several types of cells. The latter one is a problem of finding a set of chemical reactions in a given BN model of a metabolic network so that the specified (non-preferable) compounds are not producible. The result of computational experiments suggests that biologically important reactions were selected by the corresponding method. Furthermore, the method was extended and applied for handling multiple metabolic networks.

For the former one, one of the important future works is to modify the framework and methods so that real-valued expression data can be handled because gene expression time-series data are usually given as sequences of continuous values. From a theoretical viewpoint, it is important to extend Lemme 9.1 to discrimination of three or more periodic attractors. Of course, as shown in Theorem 9.1, $m(m - 1)$ genes are enough to discriminate $m$ periodic attractors. However, it seems that this value is far from optimal. Therefore, obtaining tight bounds for three or more periodic attractors is left as an open problem.

For the latter one, one of the important future works is to develop more accurate methods for handling reversible reactions because many chemical reactions in metabolic networks are reversible but *ad hoc* methods are employed for handing such reactions in the current BN models of metabolic networks. To this end, combination of the BN-based approach and the standard FBA-based approach might be useful. In addition, improvement of the computational efficiency is important future work because the exact ILP-based methods could not be applied to large metabolic networks and thus some heuristics were introduced in practical versions.

## References

1. Kauffman, S. A. (1969). Homeostasis and differentiation in random genetic control networks. *Nature* 224, pp. 177–178.
2. Akutsu, T. (2018). *Algorithms for Analysis, Inference, and Control of Boolean Networks* (World-Scientific).
3. Albert, R. and Thakar, J. (2014). Boolean modeling: A logic-based dynamic approach for understanding signaling and regulatory networks and for making useful predictions. *Wiley Interdisciplinary Reviews: Systems Biology and Medicine* 6(5), pp. 353–369.
4. Bnornholdt, S. (2008). Boolean network models of cellular regulation: Prospects and limitations. *Journal of the Royal Society Interface* 5, S1, pp. S85–S94.
5. Cheng, X., Tamura, T., Ching, W. and Akutsu, T. (2017). Discrimination of singleton and periodic attractors in Boolean networks. *Automatica* 84, pp. 205–213.
6. Tamura, T., Takemoto, K. and Akutsu, T. (2010). Finding minimum reaction cuts of metabolic networks under a boolean model using integer programming and feedback vertex sets. *International Journal of Knowledge Discovery in Bioinformatics* 1, pp. 14–31.
7. Albert, R. and Othmer, H. G. (2003). The topology of the regulatory interactions predicts the expression pattern of the segment polarity genes in drosophila melanogaster. *Journal of Theoretical Biology* 223(1), pp. 911–918.

8. Fauré, A., Naldi, A., Chaouiya, C. and Thieffry, D. (2006). Dynamical analysis of a generic boolean model for the control of the mammalian cell cycles. *Bioinformatics* 22(14), pp. e124–e131.

9. Klamt, S., Saez-Rodriguez, J., Lindquist, J. A., Simeoni, L. and Gilles, E. D. (2006). A methodology for the structural and functional analysis of signaling and regulatory networks. *BMC Bioinformatics* 7, p. 56.

10. Álvarez Silva, M. C., Yepes, S., Torres, M. M. and Barrios, A. F. G. (2015). Protein interaction network and modeling of igvh mutational status in chronic lymphocytic leukemia. *Theoretical Biology and Medical Modelling* 12, p. 12.

11. Burgard, A. P., Pharkya, P. and Maranas, C. D. (2003). Optknock: A bilevel programming framework for identifying gene knockout strategies for microbial strain optimization. *Biotechnology and Bioengineering* 84(6), pp. 647–657.

12. Orth, J. D., Thiele, I. and Palsson, B. Ø. (2010). What is flux balance analysis? *Nature Biotechnology* 28(3), pp. 245–248.

13. Jiang, D., Zhou, S. and Chen, Y.-P. P. (2009). Compensatory ability to null mutation in metabolic networks. *Biotechnology and Bioengineering* 103(2), pp. 361–369.

14. Li, Z., Wang, R.-S., Zhang, X.-S. and Chen, L. (2009). Detecting drug targets with minimum side effects in metabolic networks. *IET Systems Biology* 3(6), pp. 523–522.

15. Smart, A. G., Amaral, L. A. N. and Ottino, J. M. (2008). Cascading failure and robustness in metabolic networks. *Proceedings of the National Academy of Sciences of the USA* 105(36), pp. 13223–13228.

16. Kanehisa, M., Furumichi, M., Tanabe, M., Sato, Y. and Morishima, K. (2017). KEGG: New perspectives on genomes, pathways, diseases and drugs. *Nucleic Acids Research* 45, D1, pp. D353–D361.

17. Lu, W., Tamura, T., Song, J. and Akutsu, T. (2014). Integer programming-based method for designing synthetic metabolic networks by minimum reaction insertion in a boolean model. *PLoS ONE* 9, p. e92637.

18. Lu, W., Tamura, T., Song, J. and Akutsu, T. (2015). Computing smallest intervention strategies for multiple metabolic networks in a boolean model. *Journal of Computational Biology* 22, pp. 85–110.

19. Tamura, T., Lu, W., Song, J. and Akutsu, T. (2018). Computing minimum reaction modifications in a boolean metabolic network. *IEEE/ACM Transactions on Computational Biology and Bioinformatics* 15(6), pp. 1853–1862.

Chapter 10

# Tensor Decomposition Based Unsupervised Feature Extraction Applied to Bioinformatics

Y-h. Taguchi

*Department of Physics, Chuo University, 1-13-27 Kasuga,
Bunkyo-ku, Tokyo 112-8551, Japan*

## Abstract

Although so called "large $p$ small $n$" problem is typical in bioinformatics, there are no effective feature selection methods applicable to them. In this chapter, we propose tensor decomposition based unsupervised feature extraction. The proposed method is applied to post-traumatic stress disorder medicated heart diseases and 26 non-small cell lung cancer cell lines, off-target effect of miRNA transfection, *in silico* drug discovery from gene expression, and social insects with multiple castes. In spite of the variety of targeted problems, the proposed method turn out to work pretty well.

*Keywords*: tensor decomposition, feature selection, multi-omics, gene expression, DNA methylation, cancer cell lines.

## 10.1. Introduction

Typical data structures in bioinformatics are difficult to analyze because they are "large $p$ small $n$" problems, i.e. a small number of samples with many variables. If one applies a regression analysis to this dataset, nothing is determined because the number of fitting parameters (coefficients) is larger than the number of constraints (samples). If one attempts to select variables that are distinct between multiple conditions with some statistical tests, $P$-value adjustments considering multiple conditions often reject all positives. If supervised learning is applied to this dataset, overfitting often can occur. As a result, we usually need to reduce the degrees of freedom to proceed with further analyses.

There are multiple ways to reduce degrees of freedom. Regularization (sparse modeling), that penalizes the usage of additional variables is one such example. The disadvantage of regularization is that we must select the values of parameters that balance the prediction accuracy and the number of variables. The selection must often be performed by applying additional conditions (e.g. minimization of cross-validation errors), which are usually not decided uniquely. On the other hand, embedding methods such as principal component analysis (PCA) are often used to generate a smaller number of variables by the linear combination of original variables. The problem with this approach is that linear combination of many variables (often as many as several thousands) often can prevent us from interpreting newly generated variables.

In this chapter, I introduce a fully unsupervised methodology that is suitable for "large $p$ small $n$" problems, which is tensor decomposition (TD)-based unsupervised feature extraction (FE), that can select smaller number of variables effectively and stably. After introducing this methodology with a brief introduction of tensors and TDs, the method can be extended to treat multi-omics datasets. The method is applied to various biological problems, e.g. inference of disease-causing genes, identification of biomarkers, and *in silico* drug design considering gene expression profiles in an unsupervised manner.

## 10.2.  Tensors

Tensors are natural extensions of a matrix, in which only two kinds of variables are present (row and column variables), to more than two kinds of variables. A tensor associated with $m$ kinds of variables is denoted as an $m$-mode tensor. The components in an $m$-mode tensor are defined as

$$x_{i_1 i_2 \cdots i_m} \in \mathbb{R}^{N_1, \times \cdots \times N_\alpha \times \cdots \times N_m} \tag{10.1}$$

where $N_\alpha$, $(1 \leq \alpha \leq m)$ is the number of variables in the $\alpha$th kind of variable.

Tensors are often treated in matrix form by merging $m - 1$ kinds of variables into one kind (unfolding). In this representation, all kinds of variables other than one variable are merged as row variables, while one variable is considered as the column variable. For example, if we have a three-mode tensor $x_{ijk} \in \mathbb{R}^{2 \times 2 \times 2}$, one matrix representation is

$$\begin{pmatrix} x_{111} & x_{211} \\ x_{112} & x_{212} \\ x_{121} & x_{221} \\ x_{122} & x_{222} \end{pmatrix} \tag{10.2}$$

Unfolding often makes it easier to apply standard matrix manipulations to tensors.

## 10.3. Tensor Decomposition

TD [1] is a natural extension of matrix factorizations, especially in singular value decomposition (SVD). In SVD, the matrix is represented as a diagonal matrix multiplied by two orthogonal matrices from both sides. This makes the interpretation of the matrix easier, as it can be represented by orthogonal vectors that represent row and column variables and a diagonal matrix whose diagonal elements represent intensity (contribution) of individual vectors in orthogonal matrices.

TD is not so straightforward as SVD, because there are no representations of the combination of diagonal tensors and orthogonal matrices that can represent arbitrary tensors. Although it is possible to propose the straight extension of SVD to tensor form,

$$x_{i_1 i_2 \cdots i_m} = \sum_{\ell=1}^{L} \lambda_\ell \prod_{\alpha=1}^{m} u_{\ell i_\alpha} \tag{10.3}$$

which is known as CP decomposition [1], where $\lambda_\ell$ is a weight and $u_{\ell i_\alpha} \in \mathbb{R}^{L \times N_\alpha}$, it has not been proven that CP decomposition can approximate an $m$-mode tensor in an effective manner. For example, although the number of terms of the right-hand side, $L$, is expected to be, at most, equal to the largest number of variables among $m$ kinds of variables, i.e. $\max(\{N_\alpha\})$, there is no proof of this. Because the number of $x_{i_1 i_2 \ldots i_m}$ is $\prod_\alpha N_\alpha$, while the number of variables in the right-hand side is $L \times (\sum_\alpha N_\alpha)$, the latter is obviously much smaller than the former if $L$ is taken to be $\max(\{N_\alpha\})$. For example, when $m = 3$ and $N_\alpha = 10$, $\prod_\alpha N_\alpha = 10^3 = 1000$ while $L \times (\sum_\alpha N_\alpha) = 10 \times (10 + 10 + 10) = 300$ when $L = \max(\{N_\alpha\}) = 10$.

To address this problem, there is another extension of SVD toward tensors: the Tucker decomposition [1],

$$x_{i_1 i_2, \ldots, i_m} = \sum_{\ell_1=1}^{N_1} \cdots \sum_{\ell_\alpha=1}^{N_\alpha} \cdots \sum_{\ell_m=1}^{N_m} G(\ell_1, \ldots, \ell_\alpha, \ldots, \ell_m) \prod_\alpha u_{\ell_\alpha i_\alpha} \quad (10.4)$$

where $G \in \mathbb{R}^{N_1 \times \ldots \times N_\alpha \times \ldots N_m}$ and $u_{\ell_m i_m} \in \mathbb{R}^{N_m \times N_m}$, which is orthogonal. Although Tucker decomposition is proven to be always possible, there is another problem; it is over complete. Thus, as a consequence, it does not provide a unique representation. In actuality, we may use any orthogonal matrix, $R$, that satisfies $R^T R = I$, where $I$ is a unit matrix, $\sum_{\ell_\alpha} G(\ell_1, \ldots, \ell_\alpha, \ldots, \ell_m) u_{\ell_\alpha i_\alpha}$ in Eq. (10.4) can be rewritten as

$$\sum_{\ell_\alpha} G(\ell_1, \ldots, \ell_\alpha, \ldots, \ell_m) u_{\ell_\alpha i_\alpha} \quad (10.5)$$

$$= \sum_{\ell_\alpha, \ell_{\alpha'}, \ell_{\alpha''}} G(\ell_1, \ldots, \ell_\alpha, \ldots, \ell_m) R_{\ell_\alpha \ell_{\alpha'}} R_{\ell_{\alpha'} \ell_{\alpha''}} u_{\ell_{\alpha''} i_\alpha} \quad (10.6)$$

$$= \sum_{\ell_{\alpha'}} \left\{ \sum_{\ell_\alpha} G(\ell_1, \ldots, \ell_\alpha, \ldots, \ell_m) R_{\ell_\alpha \ell_{\alpha'}} \right\} \left( \sum_{\ell_{\alpha''}} R_{\ell_{\alpha'} \ell_{\alpha''}} u_{\ell_{\alpha''} i_\alpha} \right) \quad (10.7)$$

$$= \sum_{\ell_{\alpha'}} G'(\ell_1, \ldots, \ell_{\alpha'}, \ldots, \ell_m) u'_{\ell_{\alpha'} i_\alpha} \quad (10.8)$$

where $R_{\ell_{\alpha'} \ell_{\alpha''}}$ is the $\ell_{\alpha'}$th row and the $\ell_{\alpha''}$th column component of matrix $R$ and

$$G'(\ell_1, \ldots, \ell_{\alpha'}, \ldots, \ell_m) = \sum_{\ell_\alpha} G(\ell_1, \ldots, \ell_\alpha, \ldots, \ell_m) R_{\ell_\alpha \ell_{\alpha'}} \quad (10.9)$$

$$u'_{\ell_{\alpha'} i_\alpha} = \sum_{\ell_{\alpha''}} R_{\ell_{\alpha'} \ell_{\alpha''}} u_{\ell_{\alpha''} i_\alpha} \quad (10.10)$$

It is clear that $G'$ and $u'$ can be replaced with $G$ and $u$ in Eq. (10.4), which gives an alternative representation of Tucker decomposition.

Thus, to select a specific representation in Eq. (10.4), we need to select one specific algorithm to derive Eq. (10.4). In the following, I select the HOSVD algorithm to derive Eq. (10.4). The HOSVD algorithm [1] is implemented in R [2] as a function of hosvd in the package rTensor.

### 10.4. TD as a Tool for Multi-omics Data

TD also can be used to analyze a multi-omics dataset. Suppose we have two omics datasets $x_{ij}$ and $x_{kj}$ sharing sample $j$. $x_{ij}$ and $x_{kj}$ represent the $i$th and $k$th omics data of the $j$th sample. These two omics datasets can be integrated by generating a tensor $x_{ijk}$ as

$$x_{ijk} \equiv x_{ij} x_{kj} \tag{10.11}$$

Tucker decomposition using Eq. (10.4) can be applied to $x_{ijk}$ in Eq. (10.11) as well.

The advantages of this formulation are as follows. When we need to integrate mutli-omics datasets, we need to have weights to integrate them, because different multi-omics data have different sizes. The number of human genes is as many as $10^4$, while the number of human microRNA is as small as $10^3$, and that of methylation sites is in the millions. If we treat multi-omics data equally, it is obvious that the results will be governed by the largest dataset. Nevertheless, in our formulation of Eq. (10.11), we do not have to consider weight, as all of $x_{ijk}$ has the same amount of multi-omics data because of the multiplications performed. On the other hand, there are also disadvantages. Equation (10.11) increases the number of variables. If $x_{ij} \in \mathbb{R}^{N \times M}$ and $x_{kj} \in \mathbb{R}^{K \times M}$, the number of variables is $(N + K)M$, while Eq. (10.11) has up to $NMK$ variables. Usually, because $NK \gg N + K$, the number of variables becomes much larger.

To compensate for this disadvantage, we can reduce $x_{ijk}$ by taking the summation of $j$ as

$$x_{ik} \equiv \sum_j x_{ijk} \tag{10.12}$$

$u_{\ell k}$ and $u_{\ell i}$ can be obtained by applying SVD to $x_{ik}$. In this case, although we cannot obtain $u_{\ell j}$, this can be approximated as

$$u_{\ell j}^{(i)} = \sum_i x_{ij} u_{\ell i} \qquad (10.13)$$

$$u_{\ell j}^{(k)} = \sum_k x_{kj} u_{\ell k} \qquad (10.14)$$

Thus, we cannot avoid having two kinds of $u_{\ell j}$ in this approximation.

In the above, we assume that two omics datasets share the sample $j$. In contrast, we can integrate two populations sharing the same omics measurements, $x_{ik}$ and $x_{ij}$, that represent the $i$th omics measurement for the $k$th and $j$th samples, respectively. $x_{ijk}$ can be obtained by

$$x_{ijk} \equiv x_{ij} x_{ik} \qquad (10.15)$$

Reduction of the number of elements by partial summation is also possible using

$$x_{jk} \equiv \sum_i x_{ijk} \qquad (10.16)$$

$u_{\ell j}$ and $u_{\ell k}$ can be obtained by applying SVD to $x_{jk}$. $u s_{\ell i}$ can be defined in two ways:

$$u_{\ell i}^{(j)} = \sum_j x_{ij} u_{\ell j} \qquad (10.17)$$

$$u_{\ell i}^{(k)} = \sum_k x_{ik} u_{\ell k} \qquad (10.18)$$

All of the above relationships can be used for multi-omics data analysis.

## 10.5.  Feature Selection

Finally, I describe how to make use of the above results to select genes. Suppose that $x_{ijk}$ is the $i$th gene expression profile of the $j$th sample under the $k$th condition. For example, $j$ represents a person and $k$ represents tissue. Applying HOSVD to $x_{ijk}$, we can perform Tucker decomposition using Eq. (10.4).

Next, we inspect $u_{\ell k}$ and $u_{\ell j}$ to select those features associated with target features, e.g. "expressive in the patients' heart". Suppose that we find that $\{u_{\ell k} | \ell \in \Omega_k\}$ and $\{u_{\ell j} | \ell \in \Omega_j\}$ are associated with targeted features. Then, we list $G(\ell_i, \ell_j, \ell_k)$ with $\ell_k \in \Omega_k$ and $\ell_j \in \Omega_j$ in the order of absolute values of $G$. Finally, we can identify the top-ranked $u_{\ell i}$s whose $\ell_i$s are associated with the listed $G$s.

Those identified $u_{\ell i}$s are then used for the selection of $i$ (genes). As we would like to find $i$s which have higher contribution to $u_{\ell i}$s, we assume that $u_{\ell i}$s obey a Gaussian distribution and select $i$s associated with adjusted $P$-values that are less than a threshold (e.g. 0.01). For this, we employ $\chi^2$ distributions as

$$P_i = P_{\chi^2} \left[ > \sum_{\ell_i} \left( \frac{u_{\ell_i i}}{\sigma_{\ell_i}} \right)^2 \right] \tag{10.19}$$

where $P_{\chi^2}[> x]$ is the cumulative probability of the $\chi^2$ distribution whose argument is larger than $x$. $\sigma_{\ell_i}$ is the standard deviation. $P_i$s attributed to genes ($i$s) are further adjusted by the Benjamini–Hochberg (BH) criterion [3]. Then, genes associated with adjusted $P$-values less than 0.01 are selected.

Although I employed a three-mode tensor for the explanation, extension to more and larger mode tensors should be straightforward.

## 10.6. Synthetic Examples

To demonstrate the usefulness of TD-based unsupervised FE, we provide two synthetic examples. Both aim to analyze multi-omics data. The first one is a multi-omics tensor directly observed, while the second is a multi-omics tensor merged by multiplying two matrices observed independently.

In the first synthetic data [4], 20 cell lines ($1 \leq j \leq 20$) for which five omics data ($1 \leq k \leq 5$) are observed are assumed to exist. Thus, in total, $5 \times 20 = 100$ profiles are assumed to be available. Each profile is assumed to be composed of $10^4$ genes. Thus, the tensor is $x_{ijk} \in \mathbb{R}^{10^4 \times 20 \times 5}$. $x_{ijk} \sim \mathcal{N}_1 = \mathcal{N}(1, 1)$ or $x_{ijk} \sim \mathcal{N}_0 = \mathcal{N}(0, 1)$ is dependent upon $i, j, k$. $\mathcal{N}(\mu, \sigma)$ is a Gaussian distribution of mean $\mu$ and standard deviation $\sigma$. $x_{ijk} \sim \mathcal{N}_1$ if $i \leq 100$, or $0 < i - 100j \leq 100$, or $0 < i - (500 +$

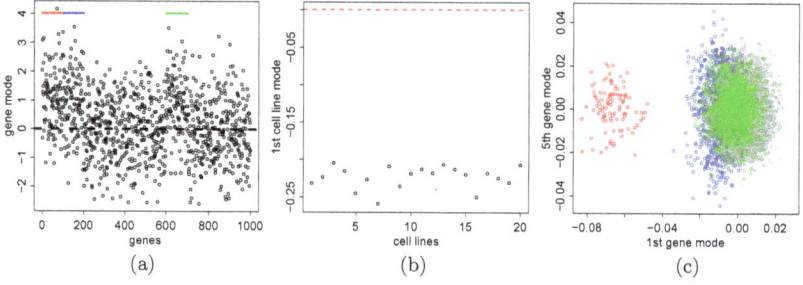

Fig. 10.1.   Artificial data. (a) $x_{ijk}$, $1 \le i \le 1000$, $j = k = 1$. One hundred genes under red segment ($1 \le i \le 100$) are always expressed. Another 100 genes under the blue segment ($101 \le i \le 200$) are expressed omics-specifically (only in one of five omics). The last 100 genes under the green segment ($601 \le i \le 700$) are expressed cell line-specifically (only in one of 20 cell lines). (b) The first singular vector, $u_{\ell_2 j}$ with $\ell_2 = 1$, cell line mode attributed to 20 cell lines. (c) A 2D scatter plot of the first and fifth gene mode, by means of singular vectors, $u_{\ell_1 i}$ with $\ell_1 = 1$(horizontal), 5(vertical). One hundred red (horizontal-coordinates $\le -0.04$), 500 blue, and 2,000 green open circles include genes under corresponding color segments in (a).

$100k) \le 100$, otherwise $x_{ijk} \sim \mathcal{N}_0$. Thus, $x_{ijk}$s with $i \le 100$ are always (regardless of cell lines or omics) expressive. Additionally, 200 genes in each profile are expressive dependent upon $j$ (cell lines) and $k$ (omics). No other genes are expressive. Figure 10.1(a) illustrates a typical example of $x_{ijk}$. The purpose of application of TD-based unsupervised FE to $x_{ijk}$ is to identify $i \le 100$ genes among $10^4$ genes. Applying TD to $x_{ijk}$, we perform the Tucker decomposition

$$x_{ijk} = \sum_{\ell_1=1}^{10^4} \sum_{\ell_2=1}^{20} \sum_{\ell_3=1}^{5} G(\ell_1, \ell_2, \ell_3) u_{\ell_1 i} u_{\ell_2 j} u_{\ell_3 k} \qquad (10.20)$$

Initially, we seek $u_{\ell_2 j}$, which lacks $j$ dependence, as it corresponds to the genes expressive regardless of 20 cell lines. As can be seen in Fig. 10.1(b), $u_{\ell_2 j}$ with $\ell_2 = 1$ does not have dependence upon $j$. Thus, it must correspond to the genes expressive regardless of cell lines. Next, we should find which $u_{\ell_3 k}$ lacks omics dependence. Nevertheless, in this specific case, we prefer to see which $G(\ell_1, \ell_2, \ell_3)$ with $\ell_2 = 1$ has larger absolute value, because this analysis is much more straightforward. To do this, we order $G$ in descending order (Table 10.1).

Table 10.1. Core tensor $G(\ell_1, \ell_2, \ell_3)$ obtained from artificial data ranked in descending order of absolute values. $\ell_1$: Gene mode, $\ell_2$: Cell line mode, $\ell_3$: Omics mode.

| Rank | $\ell_1$ | $\ell_2$ | $\ell_3$ | $G(\ell_1, \ell_2, \ell_3)$ |
|------|----------|----------|----------|------------------------------|
| 1 | 1 | 1 | 1 | −144.27316 |
| 2 | 5 | 1 | 4 | 90.20720 |
| 3 | 2 | 1 | 2 | −85.16630 |
| 4 | 3 | 1 | 3 | −81.82031 |
| 5 | 4 | 1 | 5 | 75.18319 |

As can be seen, $\ell_1 = 1$ has the maximum absolute value in $G$. In Fig. 10.1(c), $u_{\ell_1 i}$ with $\ell_1 = 1$ discriminates genes with $i \leq 100$ from other genes. Thus, TD-based unsupervised FE successfully identified the genes expressed regardless of omics and cell lines.

The second synthetic example [5] is integration of two omics datasets, as in Eq. (10.11). Here we assume $x_{ij}, x_{kj} \in \mathbb{R}^{1000 \times 50}$. In these datasets, $x_{ij}$s with $i \leq 50$ and $x_{kj}$ with $k \leq 50$ are ordered while others are random. Specifically,

$$x_{ij} = \frac{c}{2}\left(\frac{j}{M} - \sin\frac{\pi j}{M}\right) + (1 - c)\varepsilon_{ij} \qquad (10.21)$$

$$x_{kj} = \frac{c}{2}\left(\frac{M - j}{M} - \sin\frac{\pi j}{M}\right) + (1 - c)\varepsilon_{kj} \qquad (10.22)$$

for $1 \leq i, k \leq 50$, $x_{ij} = \varepsilon_{ij}$, $x_{kj} = \varepsilon_{kj}$ for $i, k > 50$. $\varepsilon$ obeys the uniform distribution $\in [0, 1]$. $c = 0.8$. We applied TD to $x_{ijk} = x_{ij} x_{kj}$.

Equations (10.21) and (10.22) can be regarded as the linear combination of three profiles, as shown in Fig. 10.2(a) to (c) with different weights. Figure 10.2(d) and (e) correspond to eqs. (10.21) and (10.22), respectively. Figure 10.2(f) is the scatter plot of Fig. 10.2(d) and (e), which does not show significant correlation between them. In spite of this, the first, second, and third sample singular value vectors $u_{\ell_3 j}$ are shown (i.e. $\ell_3 = 1, 2, 3$, Fig. 10.2(g) to (i)) to be able to reproduce three bases shown in Fig. 10.2(a) to (c). This suggests the superior capabilities of TD applied to the synthetic datasets.

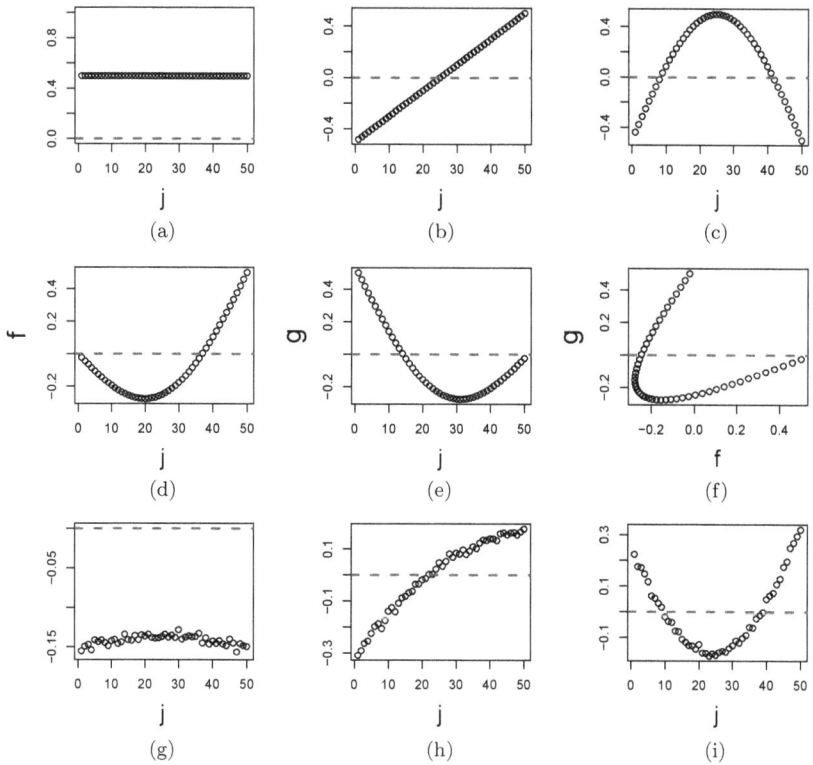

Fig. 10.2. The results of TD applied to the synthetic data [Eqs. (10.21) and (10.22)]. (a) to (c) are orthogonal base functions: (a) constant, (b) linear, (c) half-period sinusoidal, (d) and (e) are functions used for generating $x_{ij}$ and $x_{kj}$ $i, k \leq 50$. (d) $x_{ij}$ and (e) $x_{kj}$. (f) is the scatter plot of (d) and (e). (g) to (i) are the first, second, and third sample singular value vectors $u_{\ell_3 j}$ with $\ell_3 = 1, 2, 3$, respectively, and are computed by applying TD to synthetic data.

Figure 10.3(a) and (b) presents scatter plots of $u_{\ell_1 i}$ with $\ell_1 = 1, 2$ and $u_{\ell_2 k}$ with $\ell_2 = 1, 2$, respectively. It is rather obvious that $i, k \leq 50$ are successfully discriminated from others. This suggests that TD-based unsupervised FE can identify important features in an unsupervised manner.

## 10.7. Application to Real Examples

In the following, I report the applications of TD-based unsupervised FE to real data.

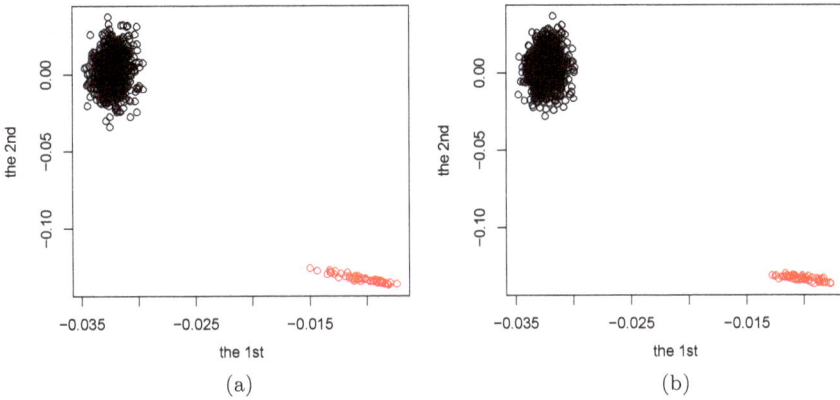

Fig. 10.3. Feature singular value vectors when TD was applied to the synthetic data [Eqs. (10.21) and (10.22)]. (a) $u_{\ell_1 i}$ with $\ell_1 = 1, 2$ and (b) $u_{\ell_2 k}$ with $\ell_2 = 1, 2$. Red dots correspond to $i, k \leq 50$.

### 10.7.1. *Post Traumatic Stress Disorder (PTSD) Mediated Heart Disease*

"Soldier's heart" [6] is a well-known disease caused by battlefield experiences that affect soldiers even after they have returned home safely. Aside from mental illnesses, soldiers often have heart problems after they return from deployment. Although the biological mechanisms are not well known, this association has been extensively studied [7]. The problem is that the heart and brain are not likely to interact with each other directly, as they are remote organs.

In this subsection, I apply TD-based unsupervised FE to gene expression in various organs of stressed mice [8]. Because gene expression must be taken from various tissues under various experimental conditions, the data inevitably must form the shape of tensors.

The form of the dataset is shown in Table 10.2. The number of organs included is 10, some of which are taken from various sub-domains of the brain. The experimental conditions are various combinations of stress and rest periods. For each combination of organs and experimental conditions, there are three to five biological replicates. They are formatted as a form of tensor, $x_{j_1 j_2 j_3 j_4 i}$, which represents the $i$th gene expression of tissues ($j_2$) under stress for duration ($j_3$) followed by a rest period after application of

Table 10.2.   Samples used in this study. Numbers before/after comma are control/treated samples.

| Stress (days) | 5 | | 10 | | | 5 | | 10 | |
|---|---|---|---|---|---|---|---|---|---|
| Rest period | 24 h | 1.5 w | 24 h | 6 w | | 24 h | 1.5 w | 24 h | 6 w |
| AY | 3,2 | 5,4 | 3,4 | 3,4 | HC | 3,5 | 4,5 | 5,4 | 4,5 |
| MPFC | 4,5 | 5,5 | 3,4 | 4,4 | SE | 3,2 | 2,3 | 3,3 | 3,3 |
| ST | 5,5 | 5,5 | 5,4 | 4,4 | VS | 5,5 | 5,5 | 3,4 | 5,4 |
| Blood | 5,5 | 5,5 | 4,5 | 4,5 | Heart | 5,5 | 4,5 | 5,5 | 5,5 |
| Hemibrain | 5,5 | 4,5 | 5,5 | 5,5 | Spleen | 5,5 | 5,5 | 5,4 | 5,5 |

*Notes*: h: hours, w: weeks, AY: amygdala, HC: hippocampus, MPFC: medial prefrontal cortex, SE: septal nucleus, ST: striatum, VS: ventral striatum.

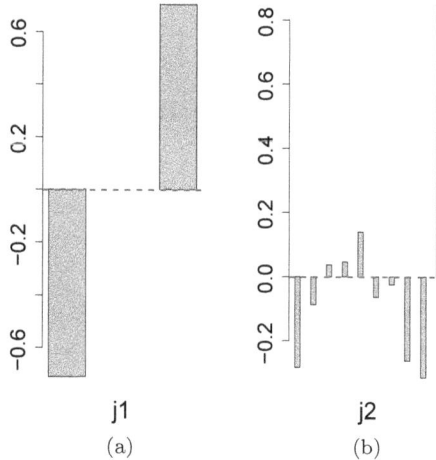

Fig. 10.4.   Singular value vectors employed. (a) The second control-related or treatment-related singular value vector, $u_{\ell_1 j_1}$, $\ell_1 = 2$. Control: $j_1 = 1$, and treatment (stress): $j_1 = 2$. (b) The fourth tissue singular value vector, $u_{\ell_2 j_2}$, $\ell_2 = 4$, AY: $j_2 = 1$, HC: $j_2 = 2$, heart: $j_2 = 8$, hemibrain: $j_2 = 9$, and spleen: $j_2 = 10$.

stress ($j_4$) with control or treatment ($j_1$) (Table 10.2). Replicates in each category were averaged before TD was applied.

Figures 10.4(a) and 10.4(b) show the second control-related or treatment-related singular value vector, $u_{\ell_1 j_1}$ with $\ell_1 = 2$ and the fourth tissue singular value vector, $u_{\ell_2 j_2}$ with $\ell_2 = 4$ obtained by TD, Eq. (10.4), applied to $x_{j_1 j_2 j_3 j_4 i}$, respectively. Figure 10.4(a) shows a clear distinction between control and treated samples while Fig. 10.4(b) shows the

Table 10.3. Top-ranked $G(\ell_1 = 2, \ell_2 = 4,$ $\ell_3, \ell_4, \ell_5)$ with greater absolute values.

| $\ell_3$ | $\ell_4$ | $\ell_5$ | $G(2, 4, \ell_3, \ell_4, \ell_5)$ |
|---|---|---|---|
| 1 | 1 | 11 | −35.0 |
| 1 | 1 | 1 | −30.8 |
| 2 | 2 | 1 | −30.3 |
| 2 | 3 | 4 | −30.0 |
| 2 | 3 | 1 | 28.7 |
| 2 | 2 | 4 | 28.5 |

simultaneous expressions among heart and brain subsections. Then, I decided to select gene singular value vectors, $u_{\ell_5 i}$ that share $G$ with larger absolute values with $u_{\ell_1 j_1}$ with $\ell_1 = 2$ and $u_{\ell_2 j_2}$ with $\ell_2 = 4$ (Table 10.3).

Table 10.3 shows the top-ranked $G$ with larger absolute values while $\ell_1 = 2$ and $\ell_2 = 4$. It is obvious that included $\ell_5$ features are restricted to 1, 4, and 11. Thus, we decided to compute $P$-values using Eq. (10.19) with $u_{\ell_5 i}$ for $\ell_5 = 1, 4, 11$. After adjusting the computed $P_i$ by the BH criterion, 801 probes were determined to be associated with adjusted $P$-values less than 0.01.

To determine whether these 801 selected probes are selectively expressive in the AY, HC, and heart as expected, the $t$-test was applied to all 40 combinations of control and treated samples. Thirteen combinations (Table 10.4) resulted in adjusted $P$-values of less than 0.01. Because the AY, HC, and heart are sufficiently represented in Table 10.4, it is clear that our strategy, TD-based unsupervised FE, successfully identified probes selectively as co-expressive in AY, HC, and heart between the control and treated samples. Thus, these 801 probes are expected to be associated with genes that cause PTSD-mediated heart diseases.

Although we have also performed extensive biological validations as well as massive performance comparisons with other methods, they are not presented here. Please see the original paper [8] for this information.

### 10.7.2. *26 Non-small Cell Lung Cancer Cell Lines*

The next example of application to a real dataset is 26 non-small cell lung cancer (NSCLC) cell lines. Although I dealt with gene expression profiles

Table 10.4.   Thirteen combinations of tissues and experimental conditions where the selected 801 probes are differentially expressed between stress-exposed and control samples.

| Stress duration | 10 days | | 5 days | |
|---|---|---|---|---|
| **Rest period** | **24 hours** | **6 weeks** | **24 hours** | **1.5 weeks** |
| AY | | ○ | | ○ |
| HC | | ○ | ○ | ○ |
| MPFC | | ○ | | |
| Heart | ○ | | | ○ |
| Hemibrain | | | ○ | ○ |
| Spleen | | ○ | ○ | ○ |

*Note*: MPFC: medial prefrontal cortex.

formatted into tensors in the previous section, I apply TD-based unsupervised FE to the multi-omics dataset [4] in this case. The datasets of TSS-seq, RNA-seq, and ChIP-seq (H3K27ac) were downloaded from DBTSS [9]. They are composed of 26 NSCLC cell lines measured genome-wide. All three datasets were summed within 25,000 bp intervals. $x_{ijk}$ should represent the $k$th omics data of the $j$th cell line at the $i$th interval. TD was separately applied to $x_{ijk}$ and computed for each chromosome.

The purpose of this analysis is as follows. Basically, the purpose of the generation of cancer cell lines is to produce a "sandbag" to test various treatments toward cancers. In contrast to expensive and time consuming *in vivo* studies, *in vitro* studies using immortal cell lines is easy and inexpensive. Thus, cancer cell lines are usually compared between treated and untreated cells. On the other hand, the characteristics of cancer cell lines themselves are not easy to determine because no normal control to be compared with cancer cell lines exists. No normal cells can ever be immortal cell lines, because immortal cells cannot be normal by definition. Because there are no cell lines to be compared with, cancer cell lines cannot be characterized easily.

Nevertheless, an unsupervised method can overcome this difficulty. By definition, unsupervised analysis can be performed without reference, because unsupervised analysis does not have to be based upon any

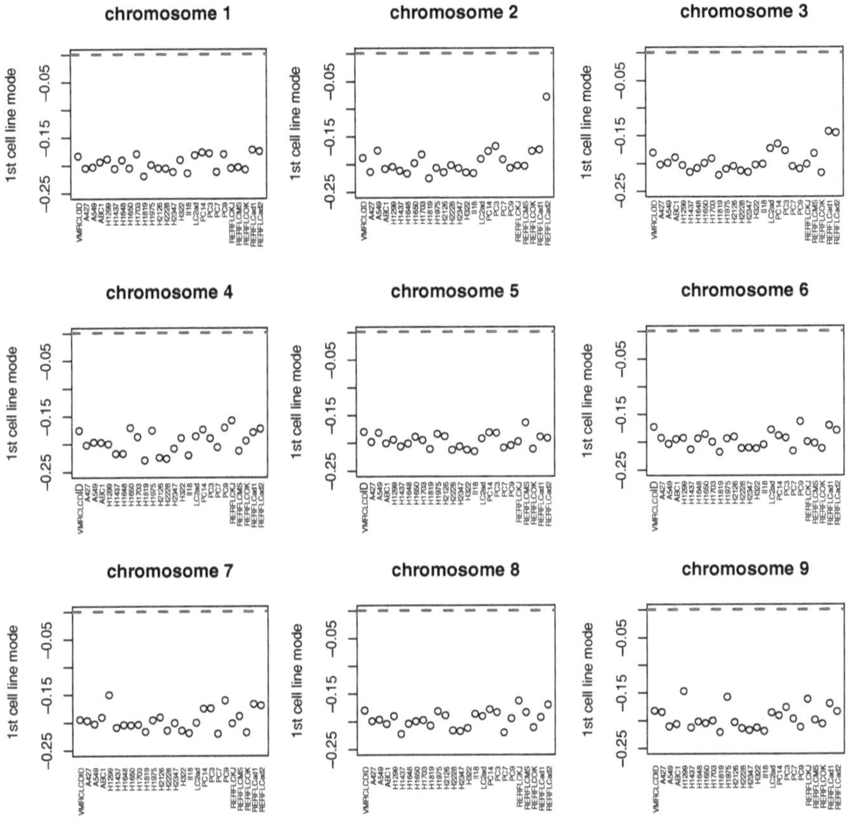

Fig. 10.5. The first singular value vectors of cell line mode $u_{\ell_2 j}$ with $\ell_2 = 1$ obtained from the 1st 9th chromosomes.

comparisons. As demonstrated in the application to synthetic data, TD-based unsupervised FE can identify genes expressive regardless of cell lines [Fig. 10.1(b)]. Such genes can be regarded as characterizing NSCLC cell lines, because they are supposed to be expressive commonly over 26 NSCLC cell lines.

Actually, regardless of the chromosomes analyzed, the first singular value vectors of cell line mode, $u_{\ell_2 j}$ with $\ell_2 = 1$, universally exhibits cell line independent profiles (Figs. 10.5, 10.6, and 10.7). Then I decided to select the interval singular value vectors, $u_{\ell_1 i}$, associated with the first singular value vectors of cell line mode $u_{\ell_2 j}$ with $\ell_2 = 1$. As a result,

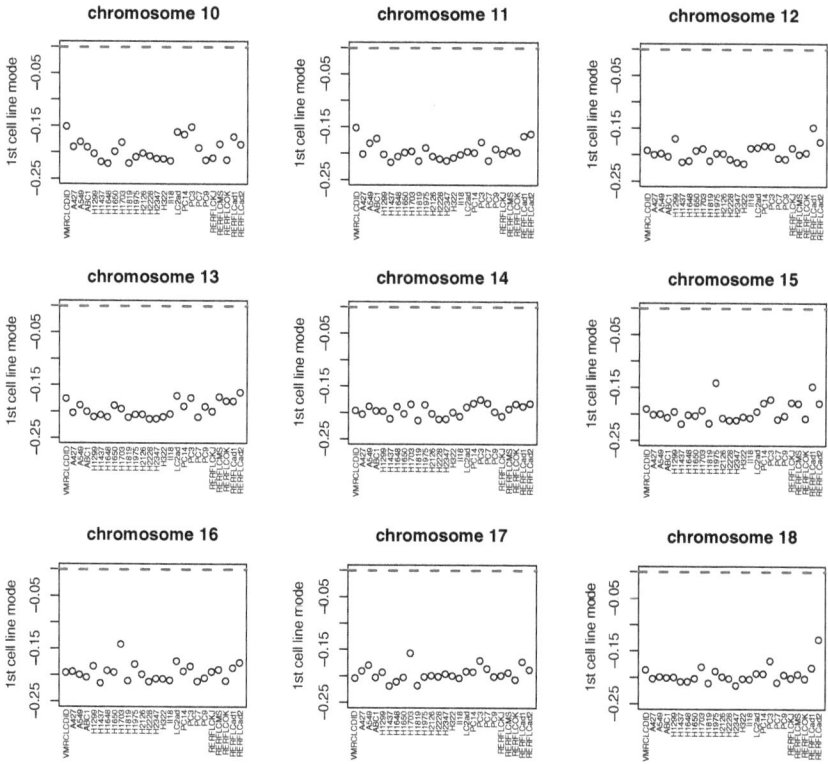

Fig. 10.6.  The first singular value vectors of cell line mode $u_{\ell_2 j}$ with $\ell_2 = 1$ obtained from the 10th to the 18th chromosomes.

I decided to employ the first three interval singular value vectors, $u_{\ell_1 i}$, $1 \leq \ell_1 \leq 3$, for attributing $P$-values to intervals using Eq. (10.19). $P_i$s attributed to each interval of individual chromosomes with Eq. (10.19) are collected and adjusted by the BH criterion. Then, 2,703 Entrez gene IDs turned out to be included in those intervals associated with adjusted $P$-values of less than 0.01 (For more details, see the original paper [4]). Thus, TD-based unsupervised FE has the capability to select genes commonly expressed over 26 cancer cell lines without any references, even if integrating three omics datasets.

Biological validation of these 2,703 genes is also available in the original paper [4].

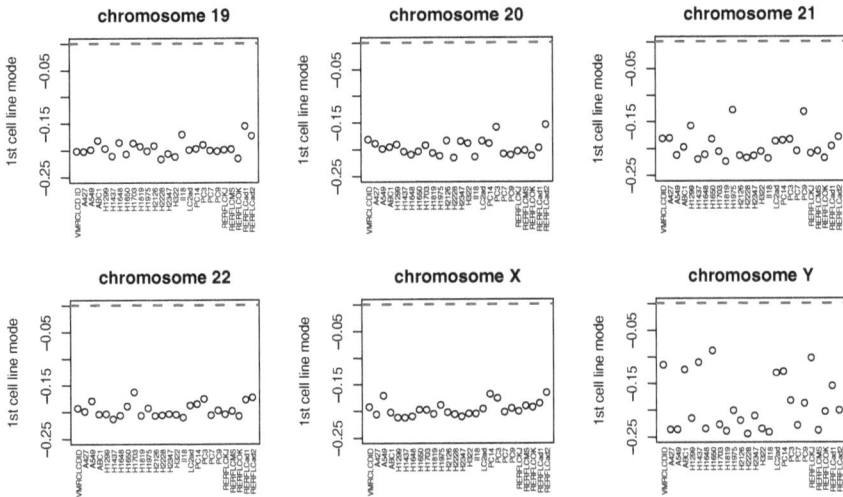

Fig. 10.7. The first singular value vectors of cell line mode $u_{\ell_2 j}$ with $\ell_2 = 1$, obtained from the 19th to the 20th, and the X, Y chromosomes.

### 10.7.3. *Universal Nature of Sequence-non-specific Off-target Regulation of mRNA Mediated by MicroRNA Transfection*

MicroRNA (miRNA) is the most well-known functional non-coding RNA, which is usually supposed to suppress target mRNAs with binding to seed mRNA regions. Although some miRNA can target hundreds of genes, their primary function is supposed to be mediated by binding to complementary regions of mRNA. In spite of this, in contrast to general belief, miRNA transfection is often associated with upregulation of various mRNAs [10], which is believed to be a side effect that lacks sequence specificity. Despite frequent observation of sequence non-specific off-target effects, it has not been studied extensively. Although numerous miRNA transfection experiments where gene expression is measured have been performed, those studies were mainly used for the investigation of canonical functions of miRNAs.

In this subsection, I apply TD-based unsupervised FE to miRNA transfection experiments [11], because TD-based unsupervised FE has the ability to identify genes expressive regardless of experimental conditions, which

Table 10.5. Eleven experiments conducted for this analysis. More detailed information is available in the original paper [11].

| Exp | GEO ID | Cell lines (cancer) | miRNA | misc |
|---|---|---|---|---|
| 1 | GSE26996 | BT549 (breast cancer) | miR-200a/b/c | |
| 2 | GSE27431 | HEY (ovarian cancer) | miR-7/128 | mas5 |
| 3 | GSE27431 | HEY (ovarian cancer) | miR-7/128 | plier |
| 4 | GSE8501 | Hela (cervical cancer) | miR-7/9/122a/ 128a/132/133a/ 142/148b/181a | |
| 5 | GSE41539 | CD1 mice | cel-miR-67, hsa-miR-590-3p, hsa-miR-199a-3p | |
| 6 | GSE93290 | multiple | miR-10a-5p,150-3p/5p, 148a-3p/5p, 499a-5p,455-3p | |
| 7 | GSE66498 | multiple | miR-205/29a/ 144-3p/5p,210, 23b,221/222/223 | |
| 8 | GSE17759 | EOC 13.31 microglia cells | miR-146a/b | (KO/OE) |
| 9 | GSE37729 | HeLa | miR-107/181b | (KO/OE) |
| 10 | GSE37729 | HEK-293 | miR-107/181b | (KO/OE) |
| 11 | GSE37729 | SH-SY5Y | 181b | (KO/OE) |

*Note*: OE: overexpression.

were cell lines in the previous section while they are transfected miRNAs in this section.

Table 10.5 summarizes the 11 experiments analyzed in this subsection. The numbers and types of transfected miRNAs differ from experiment to experiment. miRNA expression profiles are basically formatted as tensor $x_{ijk}$, which is the $i$th mRNA expression when the $j$th miRNA is transfected, while $k = 1, 2$ represents when miRNA or mock miRNA is transfected. TD was applied to $x_{ijk}$ and mRNA singular value vector, $u_{\ell_1 i}$, miRNA singular value vectors, $u_{\ell_2 j}$, and control vs treated singular value vectors $u_{\ell_3 k}$ were computed. First, we try to identify which $u_{\ell_2 j}$ lacks $j$ dependence (Usually, $\ell_2 = 1$ represents the transfected miRNA independent expression profile of mRNAs). On the other hand, $u_{\ell_3 k}$ with $\ell_3 = 2$ universally represents the distinction between treated and control samples, i.e. $u_{\ell_3 1} \simeq -u_{\ell_3 2}$ when $\ell_3 = 2$.

Which mRNA singular value vectors are used for identification of mRNAs varies from experiment to experiment (see more details in the original paper [11]). In some cases, gene expression profiles have the form of a matrix (not tensor), so in that case the computation is equivalent to standard PCA. Thus, the number of mRNAs that are selected is also highly variable (see # in Table 10.6) and varies from 100 to several hundred. Remarkably, they are highly overlapped with each other. Odds ratios are generally higher than 10 (sometimes even 100). This suggests that TD-based unsupervised FE has high capability to identify highly and commonly up/downregulated genes resulting from miRNA transfection.

The biological significance of selected variables was also evaluated (see original paper [11]) and was also found to be highly common among the 11 experiments.

The biological reason for how the sequence non-specific off-target effect takes place is not well known. One possible scenario is competition of protein machinery between endogenous and exogenous (transfected) miRNAs. If a large number of miRNAs is transfected, they can bind to protein machinery like DICER, which binds to endogenous miRNA. This results in the removal of proteins that bind to endogenous miRNAs. This process can cause a decrease of endogenous miRNAs that bind to target mRNAs. Thus, it is possible for mRNAs targeted by endogenous miRNAs to be upregulated.

To investigate the relationship between DICER and the mRNAs selected in Table 10.6, we checked if they are significantly overlapped with genes whose expression is altered by DICER KO or by overexpression as well as genes that bind to DICER (Table 10.7).

As it is obvious that identified mRNAs are tightly related to DICER, this scenario may be worth considering.

### 10.7.4. *Application to* **In Silico** *Drug Design from Gene Expression*

Because drug discovery is an expensive and time consuming process, data science was always expected to contribute to this process. There are two main streams for this direction [13]. One is ligand-based drug design (LBDD), and another is structure-based drug design (SBDD). LBDD tries to

Table 10.6.   Fisher's exact tests for coincidence among 11 miRNA transfection experiments. Upper triangle: $P$-value, lower triangle: odds ratio.

| Exp. | 1 | 2 | 3 | 4 | 5 | 6 | 7 | 8 | 9 | 10 | 11 |
|---|---|---|---|---|---|---|---|---|---|---|---|
| # | 232 | 711 | 747 | 441 | 123 | 292 | 246 | 873 | 113 | 104 | 120 |
| 1  232 |  | $4.14\times10^{-19}$ | $6.59\times10^{-22}$ | $3.96\times10^{-41}$ | $4.12\times10^{-71}$ | $9.41\times10^{-70}$ | $2.90\times10^{-60}$ | $1.34\times10^{-17}$ | $1.15\times10^{27}$ | $6.84\times10^{-26}$ | $2.66\times10^{-07}$ |
| 2  711 | 7.68 |  | 0.00 | $1.89\times10^{-18}$ | $4.93\times10^{-27}$ | $5.59\times10^{-20}$ | $2.69\times10^{-32}$ | $4.62\times10^{-13}$ | $9.23\times10^{-16}$ | $8.66\times10^{-12}$ | $1.37\times10^{-03}$ |
| 3  747 | 8.30 | 345.52 |  | $3.63\times10^{-20}$ | $7.96\times10^{-21}$ | $5.70\times10^{-12}$ | $1.82\times10^{-27}$ | $9.52\times10^{-12}$ | $1.18\times10^{-14}$ | $1.01\times10^{-12}$ | $3.90\times10^{-06}$ |
| 4  441 | 18.23 | 5.19 | 5.34 |  | $6.14\times10^{-41}$ | $1.01\times10^{-34}$ | $1.44\times10^{-69}$ | $4.61\times10^{-11}$ | $2.16\times10^{-30}$ | $4.09\times10^{-28}$ | $1.35\times10^{-10}$ |
| 5  123 | 53.86 | 9.04 | 7.27 | 17.48 |  | $2.9\times10^{-179}$ | $1.27\times10^{-63}$ | $6.24\times10^{-15}$ | $3.16\times10^{-25}$ | $2.37\times10^{-17}$ | $4.69\times10^{-09}$ |
| 6  292 | 61.50 | 8.15 | 5.52 | 17.71 | 204.39 |  | $3.53\times10^{-53}$ | $2.57\times10^{-15}$ | $6.65\times10^{-22}$ | $1.65\times10^{-12}$ | $5.60\times10^{-05}$ |
| 7  246 | 20.27 | 5.35 | 4.67 | 12.39 | 20.11 | 22.03 |  | $6.91\times10^{-42}$ | $1.77\times10^{-36}$ | $4.50\times10^{-31}$ | $2.78\times10^{-14}$ |
| 8  873 | 18.61 | 7.22 | 6.51 | 8.29 | 15.61 | 18.53 | 20.73 |  | $1.81\times10^{-07}$ | $1.37\times10^{-06}$ | $2.76\times10^{-02}$ |
| 9  113 | 39.34 | 9.87 | 8.77 | 25.98 | 32.44 | 34.90 | 21.94 | 16.02 |  | $3.7\times10^{-125}$ | $9.27\times10^{-18}$ |
| 10  104 | 40.29 | 8.22 | 8.27 | 26.64 | 23.34 | 20.86 | 21.56 | 15.18 | 517.87 |  | $6.82\times10^{-16}$ |
| 11  120 | 10.15 | 3.19 | 4.43 | 9.19 | 11.55 | 8.11 | 8.28 | 4.92 | 19.57 | 18.70 |  |

*Notes*: #: the number of genes selected for each of the 11 experiments via TD- or PCA-based unsupervised FE.

Table 10.7.  GEO DICER KO: the number of experiments among the 16 experiments included in Enrichr [12] whose set of listed genes significantly overlapped with the set of genes identified in each of our 11 experiments. IP: Fisher's exact test for the overlap between the set of genes that bind to DICER in immunoprecipitation (IP) experiments and the set of genes selected in each of the 11 experiments.

| Experiments | | 1 | 2 | 3 | 4 | 5 | 6 |
|---|---|---|---|---|---|---|---|
| GEO | up | 12/16 | 12/16 | 12/16 | 12/16 | 14/16 | 11/16 |
| DICER KO | down | 13/16 | 12/16 | 12/16 | 13/16 | 14/16 | 10/16 |
| IP | $P$-value | $2.49\times10^{-23}$ | $7.22\times10^{-22}$ | $1.31\times10^{-17}$ | $5.55\times10^{-29}$ | $5.21\times10^{-35}$ | $1.78\times10^{-20}$ |
| | odds | 47.4 | 20.6 | 15.9 | 38.7 | 64.2 | 41.2 |

| Experiments | | 7 | 8 | 9 | 10 | 11 |
|---|---|---|---|---|---|---|
| GEO | up | 12/16 | 14/16 | 12/16 | 13/16 | 12/16 |
| DICER KO | down | 12/16 | 12/16 | 14/16 | 14/16 | 10/16 |
| IP | $P$-value | $4.72\times10^{-32}$ | $4.29\times10^{-16}$ | $2.19\times10^{-11}$ | $3.96\times10^{-10}$ | $4.64\times10^{-08}$ |
| | odds | 37.0 | 41.4 | 42.6 | 39.6 | 27.3 |

seek new drugs based on the similarity between known drug compounds and new drug compound candidates. It has relatively high prediction accuracy but is useless if there are no known drug compounds. SBDD does not have this disadvantage because it can infer new drug compound candidates based on docking simulation between proteins and compounds. On the other hand, docking simulation requires massive computational resources, so SBDD is not applicable to compound libraries containing millions of compounds that LBDD can deal with more easily.

To address the disadvantages of LBDD and SBDD, a third approach was proposed. In this approach, new candidate drug compounds are proposed based upon the similarity of gene expression when cell lines and model animals are treated with the drugs. As this approach does not require massive computational research and is not based on similarity between compounds, it can infer new drugs that lack similarity with known drugs without using massive computational resources. Nevertheless, it is useless when there are no known drugs available for comparison.

In this subsection, I propose the application of TD-based unsupervised FE to drug design [14] to infer new drug candidate compounds even when there are no known drug compounds. In this prediction, we integrate two expression profiles, one being that of drug-treated cell lines or model animals and another being that of healthy humans and human patients. These

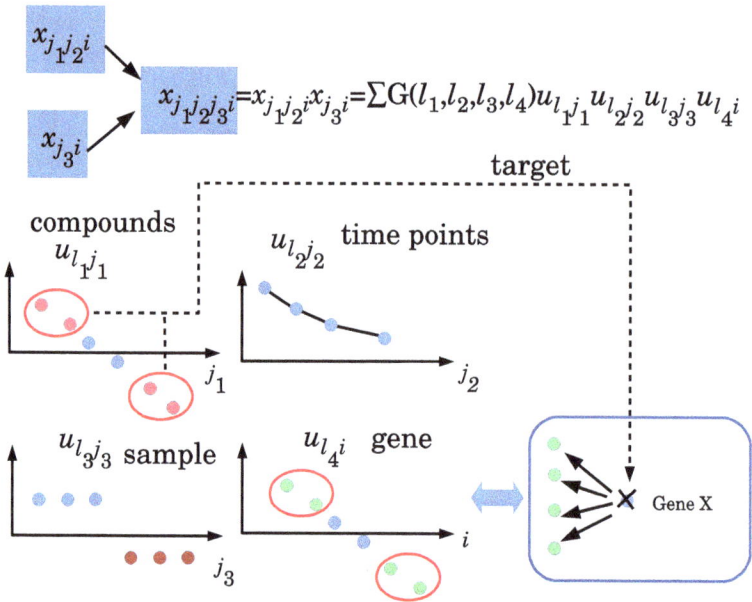

Fig. 10.8. Intuitive illustration of the present strategy. Suppose that there is a tensor, $x_{j_1 j_2 i}$, which represents the $i$th gene expression at the $j_2$th time point after the $j_1$th compound is given to a rat; these data are taken from the DrugMatrix [15] dataset. There is also a matrix, $x_{j_3 i}$, which represents the $i$th gene expression of the $j_3$th sample; samples typically include disease samples and control samples. Tensor $x_{j_1 j_2 j_3 i}$ was generated as a 'mathematical product' of $x_{j_1 j_2 i}$ and $x_{j_3 i}$ as shown in Eq. (10.15). Then, tensor $x_{j_1 j_2 j_3 i}$ is decomposed, and singular value matrix of compounds $u_{\ell_1 j_1}$, singular value matrix of time points $u_{\ell_2 j_2}$, sample singular value matrix $u_{\ell_3 j_3}$, and gene singular value matrix $u_{\ell_4 i}$ are obtained. Among them, I selected the combinations of $\ell_k$, $1 \leq k \leq 4$, which are simultaneously associated with all of the following: (i) core tensor $G(\ell_1, \ell_2, \ell_3, \ell_4)$ with a large enough absolute value, (ii) a singular value vector of time points, $u_{\ell_2 j_2}$, whose value significantly varies with time, and (iii) sample singular value vector $u_{\ell_3 j_3}$. These parameters are different between a disease (red filled circles) and control samples (cyan filled circles). Finally, using gene singular value vector $u_{\ell_4 i}$ and compound singular value vector $u_{\ell_1 j_1}$, compounds (filled pink circles) and genes (filled light-green circles) associated with $G(\ell_1, \ell_2, \ell_3, \ell_4)$s with large enough absolute values are selected. Next, if the selected genes are coincident with the genes associated with a significant alteration when gene $X$ is knocked out (or overexpressed) by Enrichr [12], then the compounds are assumed to target gene $X$.

two profiles are multiplied to generate a tensor [Eq. (10.15)] where the two profiles share genes.

Figure 10.8 illustrates how to integrate two gene expression profiles for drug design. Gene expression profiles of a model animal (Rat) treated by various compounds are downloaded from DrugMatrix [15]. Human gene expression profiles including healthy controls and patients are downloaded from the Gene Expression Omnibus (GEO) [16]. Gene expression profiles collected are as follows:

(a) Heart failure: Rat heart expression profiles treated by compounds are taken from DrugMatrix (GSE59905). Human gene expression profiles including two heart diseases and healthy controls are from GSE57345.

(b) The rat model of PTSD: Gene expression profiles of the brain for drug treatment of rats were retrieved from DrugMatrix under GEO ID GSE59895, whereas the amygdala and hippocampus gene expression for the rat model of PTSD was taken from GEO ID GSE60304.

(c) ALL (Acute Lymphoblastic Leukemia): Gene expression profiles of bone marrow for drug treatment of rats were retrieved from DrugMatrix under GEO ID GSE59894, and the ALL human bone marrow gene expression was taken from GEO ID GSE67684.

(d) Diabetes and renal cancer: Gene expression profiles of kidneys for drug treatment of rats were retrieved from DrugMatrix under GEO ID GSE59913, whereas gene expressions for diabetic human kidneys and renal cancer were obtained from GEO ID GSE30122 and GSE40435, respectively.

(e) Cirrhosis: Gene expression profiles of the liver for drug treatment of rats were retrieved from DrugMatrix under GEO ID GSE59923, whereas the gene expression for cirrhosis of the human liver was obtained from GEO ID GSE15654.

After following the procedure illustrated in Fig. 10.8, a list of proteins targeted by drugs that model a treated animal (Rat) can be obtained.

Table 10.8 summarizes the prediction performance of TD-based unsupervised FE as a drug design tool based on the gene expression profiles. Known drug targets were retrieved from DINIES [17]. It is obvious that five diseases other than ALL have predicted target proteins significantly overlapped with known proteins targeted by drugs with which model animals were treated. Thus, in conclusion, TD-based unsupervised FE has the ability to infer drug candidate compounds from gene expression profiles. For more details, see the original paper [14].

Table 10.8.  Fisher's exact test ($P_F$) and the uncorrected $\chi^2$ test ($P_{\chi^2}$) of known drug target proteins regarding the inference of the present study. Rows: Known drug target proteins (DINIES). Columns: Inferred drug target proteins using 'Single Gene Perturbations from GEO up' or 'Single Gene Perturbations from GEO down' in Enrichr [12]. OR: odds ratio.

| | | Single gene perturbations from GEO up | | | | | Single gene perturbations from GEO down | | | | |
|---|---|---|---|---|---|---|---|---|---|---|---|
| | | F | T | $P_F$ | $P_{\chi^2}$ | RO | F | T | $P_F$ | $P_{\chi^2}$ | RO |
| Heart failure | F | 521 | 517 | $3.4 \times 10^{-4}$ | $3.9 \times 10^{-4}$ | 3.02 | 628 | 416 | $1.3 \times 10^{-3}$ | $7.3 \times 10^{-4}$ | 2.61 |
| | T | 13 | 39 | | | | 19 | 33 | | | |
| PTSD | F | 500 | 560 | $3.8 \times 10^{-2}$ | $3.1 \times 10^{-2}$ | 2.67 | 532 | 529 | $6.1 \times 10^{-3}$ | $4.5 \times 10^{-3}$ | 3.81 |
| | T | 6 | 18 | | | | 5 | 19 | | | |
| ALL | F | 979 | 89 | $2.7 \times 10^{-1}$ | $3.0 \times 10^{-1}$ | 2.19 | 1009 | 57 | $1.0 \times 10^{0}$ | — | — |
| | T | 10 | 2 | | | | 12 | 0 | | | |
| Diabetes | F | 889 | 177 | $1.2 \times 10^{-2}$ | $7.1 \times 10^{-3}$ | 3.00 | 936 | 130 | $3.6 \times 10^{-4}$ | $2.0 \times 10^{-5}$ | 5.13 |
| | T | 15 | 9 | | | | 14 | 10 | | | |
| Renal carcinoma | F | 847 | 219 | $2.0 \times 10^{-2}$ | $1.2 \times 10^{-2}$ | 2.75 | 895 | 169 | $4.3 \times 10^{-2}$ | $2.2 \times 10^{-2}$ | 2.64 |
| | T | 14 | 10 | | | | 16 | 8 | | | |
| Cirrhosis | F | 572 | 219 | $1.1 \times 10^{-2}$ | $8.1 \times 10^{-3}$ | 2.91 | 595 | 169 | $1.6 \times 10^{-3}$ | $1.1 \times 10^{-3}$ | 3.81 |
| | T | 8 | 10 | | | | 7 | 8 | | | |

### 10.7.5. *Social Insects with Multiple Castes*

How a distinct phenotype appears from shared genotypes is a mystery. Basically, it can be mediated by the regulation of gene expression, although how gene expression is regulated in a phenotype dependent manner is not well known. Among possible mechanisms, epigenetic profiles are promising candidates. Thus, it would be interesting to determine if we can identify phenotype specific alterations of epigenetic profiles in coincidence with gene expression.

As a target by which the relationship between epigenome and phenotype is studied, I select multiple-caste social insects [18], e.g. bees and ants. Despite their promise, few studies have been performed on these insects. In this subsection, we attempt to identify phenotype-specific epigenetic profile alteration by applying TD-based unsupervised FE to gene expression and DNA methylation profiles of social insects that have multiple castes.

All gene expression and methylation profiles [19] were retrieved from the Gene Expression Omnibus (GEO). Gene expression profiles of *P. canadensis* and *D. quadriceps* are available as Supplementary Files in GEO ID GSE59525. Gene expression profile $x_{ij}$ of the $i$th gene at the $j$th sample and methylation profile $x_{ik}$ attributed to the $i$th gene at the $k$th sample are multiplied to generate tensor $x_{ijk}$ as in Eq. (10.15). Gene expression values were used as-is; however, methylation profile values were integrated to represent the relative methylation within genes. TD was applied to $x_{ijk}$ and we obtained gene singular value vector, $u_{\ell_1 i} \in \mathbb{R}^{N \times N}$, gene expression sample singular value vector, $u_{\ell_2 j} \in \mathbb{R}^{M \times M}$, and methylation sample singular value vector, $u_{\ell_3 k} \in \mathbb{R}^{K \times K}$, where $N$ is the number of genes, $M$ is the number of samples in which gene expression is measured, and $K$ is the number of samples in which methylation is observed.

As seen in Fig. 10.9, for *P. canadensis*, the first sample singular value vector for methylation profiles $u_{\ell_3 k}$ with $\ell_3 = 1$ (Fig. 10.9(a)) and the third sample singular value vectors for gene expression profiles, $u_{\ell_2 j}$ with $\ell_2 = 3$ (Fig. 10.9(b)) are associated with distinction among castes. After investigating core tensor $G$, the 9th and 10th gene singular value vector, $u_{\ell_1 i}$ with $9 \leq \ell_1 \leq 10$ (Fig. 10.9(c)) are selected to compute the $P$-values attributed to genes, because they share an absolutely larger $G$ with the selected sample singular vectors. Similarly, for *D. quadriceps*, the first

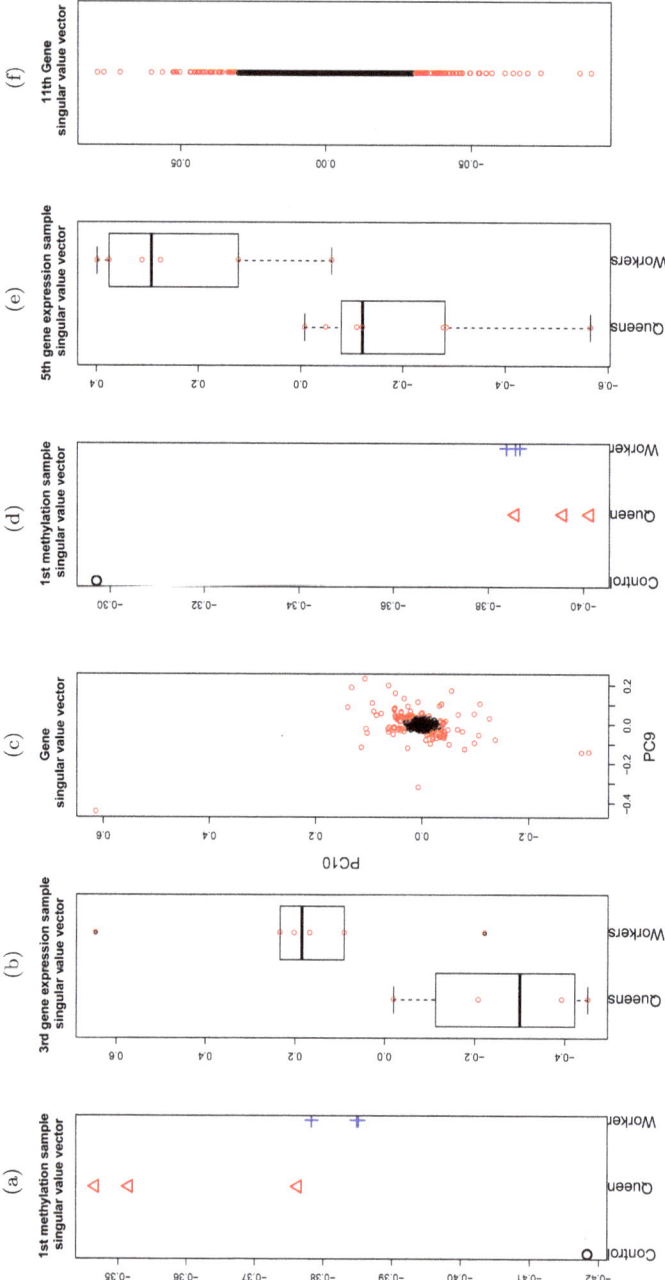

Fig. 10.9. TD analysis of gene expression profiles. TD-based unsupervised FE was applied to gene expression and methylation profiles of *P. canadensis* (a, b, c) and *D. quadriceps* (d, e, f). (a) The first sample singular value vector for methylation profiles $u_{\ell_3 k}$ (with $\ell_3 = 1$), (b) A boxplot of the third sample singular value vectors for gene expression profiles, $u_{\ell_2 j}$ (with $\ell_2 = 3$). (c) The 9th and 10th gene singular value vector, $u_{\ell_1 i}$ (with $9 \leq \ell_1 \leq 10$). Red areas in (c) represent 133 selected genes. (d) The first sample singular value vector for methylation profiles, $u_{\ell_3 k}$ (at $\ell_3 = 1$). (e) A boxplot of the fifth sample singular value vectors for gene expression profiles, $u_{\ell_2 j}$ (at $\ell_2 = 5$). (f) The 11th gene singular value vector, $u_{\ell_1 i}$ (at $\ell_1 = 11$). Red areas in (f) represent 128 selected genes.

Table 10.9. Statistical tests of the differences (between queens and workers) in gene expression and methylation. The genes identified by TD-based unsupervised FE were analyzed using $t$ (the $t$ test), Wilcox (the Wilcoxon rank sum test), and KS (the Kolmogorov–Sinai test), all two-sided.

|  |  | $t$ | Wilcox | KS |
|---|---|---|---|---|
| *P. canadensis* | gene expression | $1.71 \times 10^{-3}$ | $1.89 \times 10^{-2}$ | 0.08 |
|  | methylation | $1.74 \times 10^{-4}$ | $5.06 \times 10^{-3}$ | $1.02 \times 10^{-3}$ |
| *D. quadriceps* | gene expression | $2.73 \times 10^{-12}$ | $9.05 \times 10^{-12}$ | $4.41 \times 10^{-11}$ |
|  | methylation | 0.3757 | 0.7163 | 0.4413 |

sample singular value vector for methylation profiles, $u_{\ell_3 k}$ with $\ell_3 = 1$ (Fig. 10.9(d)) and the fifth sample singular value vectors for gene expression profiles, $u_{\ell_2 j}$ with $\ell_2 = 5$ (Fig. 10.9(e)) are associated with distinction among castes. After investigating core tensor $G$, the 11th gene singular value vector, $u_{\ell_1 i}$ with $\ell_1 = 11$ (Fig. 10.9(f)) is selected to compute the $P$-values attributed to genes, because they share absolutely larger $G$ with the selected sample singular vectors. $P$-values were attributed to genes using Eq. (10.19) with selected gene singular value vectors. As a result, 133 genes and 128 genes turned out to be associated with adjusted $P$-values (by the BH criterion) of less than 0.01 for *P. canadensis* and *D. quadriceps*, respectively.

Finally, I evaluated whether the selected genes are distinctly expressed between castes (Table 10.9). Unfortunately, methylations of *D. quadriceps* are not distinctly expressive between castes. Thus, although TD-based unsupervised FE was not fully successful, it could identify genes associated with distinct gene expression and methylation between castes simultaneously, at least for one of two species, *P. canadensis*.

Biological validation of selected genes can be found in the original paper [18].

## 10.8. Conclusions

In this chapter, I proposed the application of a TD-based unsupervised FE for gene selection. The tensors are as follows:

(i) fully filled with observed data;
(ii) generated with the multiplication of two matrices sharing samples (Eq. 10.11);

(iii) generated with the multiplication of two matrices sharing genes (Eq. 10.15).

Method (i) was applied to PTSD mediated heart disease (Section 10.7.1), non-small cell lung cancer cell lines (Section 10.7.2), and miRNA transfection (Section 10.7.3). Method (ii) was demonstrated in Section 10.6 using the second synthetic data (Although it was also applied to a real example [5], it is not included here because of space limitations). Method (iii) was applied to *in silico* drug design using gene expression (Section 10.7.4) and multiple castes of social insects (Section 10.7.5). Based on the performances achieved when TD-based unsupervised FE was applied to various biological problems, I conclude that TD-based unsupervised FE is an effective method to analyze a wide range of multi-omics datasets.

## References

1. Kolda, T. G. and Bader, B. W. (2009). Tensor decompositions and applications. *SIAM Review* 51(3), pp. 455–500, doi:10.1137/07070111x, https://doi.org/10.1137/07070111x.
2. R Core Team (2017). *R: A Language and Environment for Statistical Computing*, R Foundation for Statistical Computing, Vienna, Austria, https://www.R-project.org/.
3. Benjamini, Y. and Hochberg, Y. (1995). Controlling the false discovery rate: A practical and powerful approach to multiple testing. *Journal of the Royal Statistical Society. Series B (Methodological)* 57(1), pp. 289–300, http://www.jstor.org/stable/2346101.
4. Taguchi, Y.-H. (2017). One-class differential expression analysis using tensor decomposition-based unsupervised feature extraction applied to integrated analysis of multiple omics data from 26 lung adenocarcinoma cell lines, in *2017 IEEE 17th International Conference on Bioinformatics and Bioengineering (BIBE)*, pp. 131–138, doi:10.1109/BIBE.2017.00-66.
5. Taguchi, Y.-H. (2017). Tensor decomposition-based unsupervised feature extraction applied to matrix products for multi-view data processing. *PLoS ONE* 12(8), p. e0183933.
6. Howell, J. D. (1985). "Soldier's heart": The redefinition of heart disease and specialty formation in early twentieth-century Great Britain. *Medical History Supplement* 5, pp. 34–52.
7. Coughlin, S. S. (2011). Post-traumatic stress disorder and cardiovascular disease. *The Open Cardiovascular Medicine Journal* 5(1), pp. 164–170, doi:10.2174/1874192401105010164, https://doi.org/10.2174/1874192401105010164.
8. Taguchi, Y.-H. (2017). Tensor decomposition-based unsupervised feature extraction identifies candidate genes that induce post-traumatic stress disorder-mediated heart diseases. *BMC Medical Genomics* 10(Suppl 4), p. 67.

9. Suzuki, A., Kawano, S., Mitsuyama, T., Suyama, M., Kanai, Y., Shirahige, K., Sasaki, H., Tokunaga, K., Tsuchihara, K., Sugano, S., Nakai, K. and Suzuki, Y. (2017). DBTSS/DBKERO for integrated analysis of transcriptional regulation. *Nucleic Acids Research* 46(D1), pp. D229–D238, doi:10.1093/nar/gkx1001, https://doi.org/ 10.1093/nar/gkx1001.

10. Khan, A. A., Betel, D., Miller, M. L., Sander, C., Leslie, C. S. and Marks, D. S. (2009). Transfection of small RNAs globally perturbs gene regulation by endogenous microRNAs. *Nature Biotechnology* 27(6), pp. 549–555, doi:10.1038/nbt.1543, https: //doi.org/10.1038/nbt.1543.

11. Taguchi, Y.-H. (2018). Tensor decomposition-based unsupervised feature extrac- tion can identify the universal nature of sequence-nonspecific off-target regulation of mRNA mediated by microRNA transfection. *Cells* 7(6), p. 54, doi:10.3390/ cells7060054, http://www.mdpi.com/2073-4409/7/6/54.

12. Kuleshov, M. V., Jones, M. R., Rouillard, A. D., Fernandez, N. F., Duan, Q., Wang, Z., Koplev, S., Jenkins, S. L., Jagodnik, K. M., Lachmann, A., McDermott, M. G., Monteiro, C. D., Gundersen, G. W. and Ma'ayan, A. (2016). Enrichr: A comprehensive gene set enrichment analysis web server 2016 update. *Nucleic Acids Research* 44(W1), pp. W90–97.

13. Cavasotto, C. (ed.) (2015). *In Silico Drug Discovery and Design* (CRC Press), doi: 10.1201/b18799, https://doi.org/10.1201/b18799.

14. Taguchi, Y.-H. (2017). Identification of candidate drugs using tensor-decomposition- based unsupervised feature extraction in integrated analysis of gene expression between diseases and drugmatrix datasets. *Scientific Reports* 7(1), p. 13733.

15. National Toxicology Program (2010). DrugMatrix, https://ntp.niehs.nih.gov/ drugmatrix/index.html.

16. Barrett, T., Wilhite, S. E., Ledoux, P., Evangelista, C., Kim, I. F., Tomashevsky, M., Marshall, K. A., Phillippy, K. H., Sherman, P. M., Holko, M., Yefanov, A., Lee, H., Zhang, N., Robertson, C. L., Serova, N., Davis, S. and Soboleva, A. (2013). NCBI GEO: Archive for functional genomics datasets-update. *Nucleic Acids Research* 41(Database issue), pp. D991–995.

17. Yamanishi, Y., Kotera, M., Moriya, Y., Sawada, R., Kanehisa, M. and Goto, S. (2014). DINIES: drug-target interaction network inference engine based on supervised anal- ysis. *Nucleic Acids Research* 42(Web Server issue), pp. 39–45.

18. Taguchi, Y.-H. (2018). Tensor decomposition/principal component analysis based unsupervised feature extraction applied to brain gene expression and methylation profiles of social insects with multiple castes. *BMC Bioinformatics* 19(Suppl 4), p. 99.

19. Patalano, S., Vlasova, A., Wyatt, C., Ewels, P., Camara, F., Ferreira, P. G., Asher, C. L., Jurkowski, T. P., Segonds-Pichon, A., Bachman, M., González-Navarrete, I., Minoche, A. E., Krueger, F., Lowy, E., Marcet-Houben, M., Rodriguez-Ales, J. L., Nascimento, F. S., Balasubramanian, S., Gabaldon, T., Tarver, J. E., Andrews, S., Himmelbauer, H., Hughes, W. O. H., Guigó, R., Reik, W. and Sumner, S. (2015). Molecular signatures of plastic phenotypes in two eusocial insect species with simple societies. *Proceedings of the National Academy of Sciences* 112(45), pp. 13970–13975, doi:10.1073/pnas. 1515937112, https://doi.org/10.1073/pnas.1515937112.

# Index

**A**

acquired resistance, 49
Acute Lymphoblastic Leukemia, 181
Apache Hadoop, 109, 112
Apache Spark, 116
apoptotic, 50
artificial intelligence, 1
attractors, 142–144, 146–148, 156
autism, 12
autism case studies, 13
autism etiology, 12
autism genetics, 24
autism subgroups, 13

**B**

big data, 15, 59, 73
bioinformatics, 11
blockchain adapter, 30, 38, 41
blockchain adapter design guidelines, 40
blockchain-centric architecture, 30, 42
bloodstream, 50
Boolean functions, 141–142
Boolean networks, 141–142, 156
bootstrapping method, 89, 94, 96–97, 103

**C**

$\chi^2$ distribution, 165
cancer gene signatures, 109
cancers, 47
caste, 185
cell-free DNA (cfDNA), 50

cell-free miRNA, 50
cell-free mRNA, 50
cell-free nucleic acid, 50
Chinese Remainder Theorem, 147
ChIP-seq, 172
circulating cancer genome atlas (CCGA)
    study, 53
circulating tumor DNA (ctDNA), 47
cluster analysis, 88, 90–91, 94–95, 103
cluster validity, 97, 101
Common Data Element (*see also* CDE),
    29, 33
common genetic variation, 12
complex disease, 20
complex disease etiology, 12, 25
compound nodes, 149–151, 153
Contrast Mining, 16
Cox proportional hazards model, 1
Cox regression, 1
CP decomposition, 161
cross-validation, 160
ctDNA analysis in directing treatment
    option, 54

**D**

data mining, 13
data-driven, 25
degrees of freedom, 160
detecting primary cancer, 54
detecting recurrence, 54
determining the prognosis of patients,
    54

development of drug resistance of cancers, 50
diabetes, 181
Dindex, 92, 101–102
disease etiology, 21
diseases, 11
distance-based, 95, 103
distance-based clustering, 89
drug design, 59, 61, 71
DrugMatrix, 181
dynamic programming, 145

**E**

EGFR amplification, 49
EGFR TKI erlotinib or gefitinib, 49
$EGFR^{L858R}$ (mutant allele fraction: 15%) and $EGFR^{T790M}$ (mutant allele fraction: 8.5%), 48
EM algorithm, 92
emotional and behavioral problems, 13
Encrypted Data Storage server, 39
enrichment analysis, 22
epigenetic, 183
epigenome, 183
Ethereum Blockchain, 30, 38
evolutionary forces, 50
exocytotic cells, 50
expectation–maximization, 92, 96
exploratory analysis, 25

**F**

F-statistics, 1
feasible solution, 152
feature extraction, 160
feature selection, 164
feature weights, 94, 96–97
Fisher's exact test, 179
flux balance analysis (FBA), 149, 157
Frequent Pattern Mining, 15
functional classification, 24

**G**

gene annotations, 95
gene expression, 88, 91–92, 144

Gene Expression Omnibus, 183
gene functional similarity, 89, 93, 103
gene interactions, 19
gene ontology, 93
gene pairs, 19, 26
gene regulatory network, 144
gene semantic similarities, 96
generalized iterative modeling, 1
generalized linear models, 1
genetic association, 17
genetic distinctions, 16
genetic drift, 50
genetic patterns, 15, 25
genetic variants, 15
genomics, 1–2, 49
GitHub, 1
Global User ID, 35
glycomics, 1–2

**H**

HAC, 101–103
heritable diseases, 26
Heritable Genotype Contrast Mining, 13
HOSVD, 163
Hubert gamma statistics, 92, 101–103
hypergeometric distribution, 17

**I**

IBIS and BRICS, 29–30
IBIS central authentication server, 35
IBIS data collection, 34
IBIS data dictionary, 35
IBIS data repository, 34–35
IBIS form structure and e-form, 33
IBIS GUID, 35
IBIS meta study, 35
IBIS query tool, 35
IBIS submission tool, 34
IBIS/BRICS dataflow, 33
IBIS/BRICS dictionary module, 33
immune system, 24
inherited risk factors, 16
*in silico* drug design, 160

integer linear programming (ILP), 149,
  152–155, 157
Integrated Biomedical Informatics
  System, 32
intra-tumoral heterogeneity, 51
*in vitro* studies, 172
*in vivo* studies, 172
IVDMIA, 111

**K**

*k*-means, 88, 95–97, 101, 103–104
KEGG, 155–156
KL divergence, 92–93

**L**

language skills, 13
ligand-based drug design, 177
linear integer programming, 149
linear programming (LP), 152–153

**M**

MapReduce, 116
maximal valid assignment, 154
metabolic networks, 141–142, 149–152,
  154–157
metabolic pathways, 155
metabolomics, 1–2
methylation, 163
microarray data, 88
microRNA, 163
migration, 50
Mirth Connect, 30, 37
mixed integer linear programming
  (MILP), 153
model expressibility, 1
model-based clustering, 91
model-based gene pair distance, 89,
  91–93
model-based methods, 88
model-view-controller, 32
molecular docking, 61, 63–65, 69–70, 74,
  80
molecular dynamics, 61, 63, 69, 71, 73,
  76, 80

multi-omics, 160
mutant allele fraction (MAF), 51
mutation, 50

**N**

natural products, 60–61, 63, 80
NB distributions, 88–89, 91
necrotic, 50
next-generation sequencing, 87
non-coding RNA, 175
non-invasive blood draw, 54
non-small cell lung cancer, 49, 171
normalized mutual information (NMI),
  96–100
NP-hard, 145, 152–153

**O**

Odds ratio, 177
optimal solution, 152
orthogonal partial least squares
  discriminant analysis, 2
osimertinib, 52

**P**

pathway enrichment analysis, 17,
  19–20
pathway pairs, 19
pathway priority, 21–22
PCR and NGS technologies, 52
Pearson's coefficient, 95
periodic attractor, 143–144, 146–149,
  157
phenotype, 183
Poisson distribution, 91, 96–97
polynomial time, 152–153
post traumatic stress disorder, 169
principal component analysis, 160
prioritized pathways, 26
progression, 50
proteomics, 1–2

**R**

reaction cut, 142, 150–152, 154–156
reaction nodes, 149–151, 153

regularization, 160
regulatory networks, 88, 104
renal cancer, 181
RNA-seq, 87–92, 95, 103, 111, 172

**S**

selection, 50
semantic similarity, 93
semantic-based, 91, 95, 97, 101, 103–104
semantic-based gene pair distances, 89,
 91, 93–94
sequence-non-specific off-target
 regulation, 175
serial blood sampling, 48
shape initiation, 50
singleton attractor, 143–145, 147–149
singular value decomposition, 161
smart contract, data manager, 41
smart contract, data store, 41
smart contract, identity management, 42
smart contracts, 38
Smart Contracts State and Function
 Details, 41
SOM, 95–97, 101, 103–104
somatic genomic alteration, 47
source compounds, 150
source nodes, 150
sparse modeling, 160

Spring Security, 35
state transition diagram, 143
statistical methods, 17
structure-based drug design, 177
supervised learning, 160
survival analysis, 1
synaptic function, 24

**T**

$t$-test, 171
target compounds, 155
target nodes, 150
technological highlight of ctDNA
 analysis, 48
tensor decomposition, 160–161
transcriptomic, 1–2, 61, 63, 78–80
transformation to small cell carcinoma, 49
TSS-seq, 172
Tucker decomposition, 162

**U**

U-statistics, 1
unfolding, 161
unique combinations, 16

**V**

valid assignments, 150–151, 154

www.ingramcontent.com/pod-product-compliance
Lightning Source LLC
Chambersburg PA
CBHW050603190326
41458CB00007B/2151